# Global Ethics on Climate Change

The volatility of climate change is increasing. It is bad news, and many climatologists, policy analysts and environmental groups regard the West as the largest contributor to the problems caused by climate change. This book raises questions concerning the systemic and cultural reasons for Western countries' unwillingness to bear full responsibility for their carbon emissions. Is the Western paradigm failing? Can other cultures offer solutions? Are there alternatives for designing a better future?

Just as the roots of the problem of climate change are cultural, the solution must be too. The contributors to *Global Ethics on Climate Change* explore cultural alternatives. This differs from conventional climate ethics, which tends to address the crisis with utilitarian, legalistic, and analytic tools. The authors in this volume doubt whether such paradigm patches will work. It may be time to think outside the box and consider non-Western insights about the good life, indigenous wisdom on being-in-the-world, and new ideas for civil evolution. This book is an examination of candidates for a Plan B.

This book was published as a special issue of the *Journal of Global Ethics.*

**Martin Schönfeld** is a former Kant Scholar and Historian of the European Enlightenment. He teaches at the University of South Florida and conducts philosophy workshops in Taiwan. He organized the first International Philosophy Conference on Climate (2006) and edits the Climate Philosophy Newsletter. He writes on climate, the American Disenlightenment, and civil evolution.

# Global Ethics on Climate Change
## The Planetary Crisis and Philosophical Alternatives

*Edited by*
**Martin Schönfeld**

Routledge
Taylor & Francis Group
LONDON AND NEW YORK

First published 2013
by Routledge
2 Park Square, Milton Park, Abingdon, Oxon, OX14 4RN

Simultaneously published in the USA and Canada
by Routledge
711 Third Avenue, New York, NY 10017

*Routledge is an imprint of the Taylor & Francis Group, an informa business*



This book is a reproduction of the *Journal of Global Ethics*, vol. 7, issue 2. The Publisher requests to those authors who may be citing this book to state, also, the bibliographical details of the special issue on which the book was based.



*British Library Cataloguing in Publication Data*
A catalogue record for this book is available from the British Library

ISBN 13: 978-0-415-50918-3

Typeset in Times New Roman
by Taylor & Francis Books

**Publisher's Note**
The publisher would like to make readers aware that the chapters in this book may be referred to as articles as they are identical to the articles published in the special issue. The publisher accepts responsibility for any inconsistencies that may have arisen in the course of preparing this volume for print.

# Contents

# Citation Information

The following chapters were originally published in the *Journal of Global Ethics*, volume 7, issue 2 (August 2011). When citing this material, please use the original page numbering for each article, as follows:

**Chapter 1**
*Introduction: Plan B: global ethics on climate change*
Martin Schönfeld
*Journal of Global Ethics*, volume 7, issue 2 (August 2011) pp. 129–136

**Chapter 2**
*Climate, imagination, Kant, and situational awareness*
Michael Thompson
*Journal of Global Ethics*, volume 7, issue 2 (August 2011) pp. 137–148

**Chapter 3**
*Moral progress and Canada's climate failure*
Byron Williston
*Journal of Global Ethics*, volume 7, issue 2 (August 2011) pp. 149–160

**Chapter 4**
*Climate change and philosophy in Latin America*
Ernesto O. Hernández
*Journal of Global Ethics*, volume 7, issue 2 (August 2011) pp. 161–172

**Chapter 5**
*Watsuji Tetsuro, Fudo, and climate change*
Bruce B. Janz
*Journal of Global Ethics*, volume 7, issue 2 (August 2011) pp. 173–184

**Chapter 6**
*Climate change and the ecological intelligence of Confucius*
Shih-yu Kuo
*Journal of Global Ethics*, volume 7, issue 2 (August 2011) pp. 185–194

**Chapter 7**
*A Daoist response to climate change*

Chen Xia and Martin Schönfeld
*Journal of Global Ethics*, volume 7, issue 2 (August 2011) pp. 195–204

**Chapter 8**
*Justice, negative GHIs, and the consumption of farmed animal products*
Jan Deckers
*Journal of Global Ethics*, volume 7, issue 2 (August 2011) pp. 205–216

# Notes on Contributors

**Jan Deckers** is a Bioethicist in Newcastle University's Medical School. He has published on ethical theory, the ethics of abortion and embryo research, and animal ethics. His current work focusses on the ethical issues associated with the consumption of products that have been derived from the bodies of animals.

**Sirkku Hellsten** is Professor of Philosophy at the University of Dar es Salaam, Tanzania, and Adjunct Professor of Social and Moral Philosophy at the University of Helsinki, Finland. Professor Hellsten has worked also as a Counsellor for Economics, Governance and Human Rights at the Finnish Embassies in Maputo (Mozambique) and Nairobi (Kenya); before that she was Reader in Development Ethics and Director of the Centre for the Study of Global Ethics at the University of Birmingham, UK. Professor Hellsten has published extensively on the issues related to development ethics, bioethics, global ethics, social justice and human rights. She is also a member of the advisory board of the International Development Ethics Association and an Editor of the *Journal of Global Ethics.*

**Ernesto O. Hernández** is pursuing a doctoral degree at the University of South Florida, Tampa, USA. He received a BA in Philosophy from the University of Puerto Rico in 2003 and an MA in Philosophy from the University of South Florida in 2010. His research interests include ethics, Latin American thought, colonialism critique, and climate philosophy.

**Bruce B. Janz** is Chair of the Philosophy Department, Director of the Center for Humanities and Digital Research, and Associate Professor of Humanities at the University of Central Florida. He writes on African philosophy, concepts of place and space across multiple disciplines, and cultural philosophy. His book *Philosophy in an African place* was published in 2009, and he is currently working on a book called *Philosophy as if place mattered.* He has taught at universities in Canada, Kenya, South Africa, and USA.

**Shih-yu Kuo** is a Professorial Research Fellow at the Academica Sinica, Taipei, Taiwan. She works on environmental ethics, ecological cultural studies, and philosophy of education and has published numerous papers on these subjects in various Chinese journals.

**Timothy B. Leduc** completed a Master's Degree in Social Work and worked for a number of years in Northern indigenous communities and on anti-violence issues before graduating with a PhD in Environmental Studies. His doctoral research built on these experiences by bringing Inuit cultural views of Northern warming into dialogue with Western climate research as a means for assessing the current politics of inaction. The resulting book, *Climate, culture, change: Inuit and Western dialogues with a warming North*, was short-listed for the 2012 Canada Prize, and related articles have been published in *Ethics and the Environment, Climatic Change* and *Worldviews*. The working title for his current book is *A Canadian ecology of mind: From the fur trade to the tar sands and beyond*. He is an Assistant Professor of Environmental Studies at York University in Toronto, Canada.

**Frederick Ochieng'-Odhiambo** holds a PhD from the University of Nairobi, Kenya. He is Head of the Department of History and Philosophy at the University of the West Indies, Barbados, where he teaches Philosophy and is Professor of African Philosophic Sagacity. His research interest is in the area of African philosophy, social and political philosophy, and logic; areas in which he has written several books and numerous articles published in international refereed journals.

**Martin Schönfeld** is Professor of Philosophy at the University of South Florida, Tampa, USA. He works in the history of ideas and in the philosophy of climate. He organized the first International Conference on Climate and Philosophy, in 2006, at the University of South Florida, and edits the annual Climate Philosophy Newsletter. Recent publications include 'Amerigenic climate change: An indictment of normalcy', *Human Ecology Review* 17 (2010), 'Climate change and cognitive evolution', in R. Irwin, ed., *Climate change and philosophy* (Continuum, 2010), and the topic issue on climate ethics (Editor) of *Essays in philosophy* (2010).

**Michael Thompson** received his PhD from the University of South Florida. He is an Adjunct Professor at the University of Tampa. His research interests include Kant's critical philosophy, the imagination and its role in cognition, and nineteenth-century German idealism. Some recent publications are 'Personhood in Kant, Confucius, and Lao Tzu', in S. Palmquist, ed., *Cultivating personhood: Kant and Asian philosophy* (de Gruyter, 2011); 'Kant's solution to the 3rd Antinomy and its legacy', *National Taiwan University Philosophical* Papers (2010); and 'An open-ended conversation: Reconciling European and African philosophy' in Brandon et al., eds., *Conversations in philosophy: Crossing the boundaries* (Cambridge Scholars, 2008).

**Byron Williston** received his PhD in Philosophy from the University of Toronto. He is Associate Professor of Philosophy at Wilfrid Laurier University in Waterloo, Canada. His primary areas of interest are environmental ethics, moral psychology, and early-modern philosophy. He has published on Descartes, Locke, Nietzsche, motivated irrationality, the philosophy of emotions, moral reasons, and moral dilemmas, among other topics. His most recent book is *Environmental ethics for Canadians*, Oxford University Press, 2011.

**Chen Xia** is a Research Fellow at the Institute of Philosophy, Chinese Academy of Social Sciences (CASS), Beijing, China. During these years, she has been a Visiting Scholar at Harvard-Yenching Institute and SOAS of the University of London. Currently, she is a Fulbright Visiting Scholar at Brown University. Her specialty is Religions in China and Chinese Philosophy, concentrating on Daoism. She is the author of *Studies of Daoist moral tracts*, the co-chief Editor and main contributor of *Principles in the study of religions* and chief Editor for *Studies of Daoist ecological* *thoughts.* She is one of the translators for books like *Daoism and traditional Chinese culture, Martin Luther's theological thoughts, Man's religions* and *Daoism and ecology.* She has contributed articles to a number of journals.

# INTRODUCTION

## Plan B: global ethics on climate change

Martin Schönfeld

*Philosophy, University of South Florida, Tampa, FL, USA*

This introduction to the special issue on climate puts individual contributions in context. Climate change is the result of the current civilization paradigm and its modes of cognition. This suggests a failure of conventional ways of thinking, including mainstream philosophy. The articles in this issue illustrate alternative philosophical approaches, which point to civil evolution.

If we are not careful, climate change may trigger the greatest catastrophe in the history of civilization. Climate change is not a future danger anymore. The processes are well underway, and a destabilization of the Earth system has begun. This lends vital urgency to the question that is at the heart of ethics: what should we do?

Philosophy is the rational investigation of existence in the world, but the world is different now from the one our ancestors inhabited. The difference concerns not so much the obvious phenomena one associates with modernity, such as urbanization, industry, and technology, because all of them, in various ways, have shaped the world already for centuries. The real and radical difference, between our generation and all the ones in the past, is the collective arrival at the limit of existing in the world. The world is our *oikos* or house; we have filled this ecological house over time, and now the house is full. Reaching such limits is an experience bygone cultures knew as well, but only in localized form. This is the first time in all of history that global civilization without exception – the sum-total of humankind – arrives at this juncture.

Hence, being-in-the-world is qualitatively different now, and this affects its rational investigation. The climate crisis is the systematic expression of the arrival at the limit. This emerging reality is so sweeping that it creates a new frame of reference for the other core questions of philosophy as well. Each of them has to be asked anew. Thus: what can we know? What may we hope? And what, really, does it mean to be human? Or, as Heidegger reformulates these queries: where do we stand? Where are we heading?[1]

### 1. Where are we heading?

Engaging with all of these questions illuminates the human and global dimensions of the climate crisis. The last query, of *where we are heading*, asks us to survey the path we are on. If 'we' is taken to mean humankind now, and 'heading' is understood as the probability-cone of likely futures as charted by the UN IPCC Assessment Reports and supplementary studies at present, then the peer-reviewed consensus is that we are heading straight into a bottleneck of our own making. This looming chokepoint is created by two processes on a collision course. Demand for everything we need is rising. That is the one process. Supply of everything we need is falling. This is the other process. Their collision is the bottleneck.

Demand for everything that makes physical existence possible – water, food, energy, and land – is only going to go up. The world population is expected to level off after a mid-century peak of nine billion. Because of near exponential population growth until quite recently, a large share of our species has not yet reached reproductive age. The result is a massively built-up demographic momentum; even if we wanted to stop multiplying now, we could not. A high speed train, with brakes fully set, slides a while before coming to a stop. Similarly, we are incapable of curbing the increase of our numbers in civil and consensual ways *before* adding two more billion to the seven billion we are already. But the planetary surface is not going to expand with us, and neither will the biospherical services our collective existence depends on. Hence, competition for available water, food, energy, and land is fated to become stiffer.

The situation is made worse by economic structures. The collapse of communism had the effect that essentially our entire species bet its house on the market economy. It is essential for the market to grow; that is, economic stability is tied to economic expansion. And indeed, the world economy grew 19-fold between 1900 and 2000 and is only set to grow more.[2] This characteristic of the market economy is so essential, so consequential, and at the same time so bizarre that it is hard to overstate its importance: *normal* business means to do *more* business, while doing the *same* business means to do *less* business. Whenever growth slows, stops, and reverses, business is hurting and the market has difficulties sustaining social services. Thus, our well-being has come to depend on a system whose integrity requires growth.

Since no one knows how to ensure social security in a shrinking market, governments are committed to stimulate further growth. Growth means more demands on the commons. Thus, strains on the biosphere will inexorably increase. Akin to a malignancy metastasizing in an organism, the market is appropriating ever more of nature. For reasons of economic design alone, demand for everything we use must accordingly go up as well.

But since nature does *not* expand, supply must go down. It not only goes down relative to demand, but it also goes down in absolute terms. The systemic reason was best summed up by Lester Brown in 1998:

> As the economy grows, pressures on the Earth's natural systems and resources intensify . . . evidence of mounting stresses can be seen on every land as more and more sustainable yield thresholds are crossed and as waste absorptive capacities are overwhelmed. (Brown 1998, 2–3)

In fact, sustainable yield thresholds of all resources we need have been crossed in the past decade. Biological diversity is declining. World fish stocks are being used up. Rainforests are shrinking. Soil erosion is spreading. Deserts are expanding. Water tables are falling. The capacity of the Earth system to nourish us and to absorb our waste is steadily diminishing. Unsurprisingly, in an updated summary of the state of the Earth system, L. Brown in 2011 comes to the conclusion that *the world is on the edge*.[3]

Overpopulation and overconsumption overwhelm biospherical capacities. If it were not for climate change, this looming chokepoint would still be ways ahead. But enter climate change into the equation, and everything gets worse – the chokepoint is now just around the corner. Looming constraints slide toward us, from the far future to the immediate tomorrow. Before climate change, the bottleneck was a risk for future generations. Through climate change, it imposes a burden on everyone alive now.

Climate change magnifies every single empirical component of the mounting stress on the Earth system – it speeds up biodiversity loss, destabilizes the ocean food-chain, converts jungle to savannah, worsens soil erosion, boosts desertification, and dries out the land. It does so through two intertwined developments, *global warming* and what for want of a better term might be called a *power surge*.

Global warming adversely affects water cycles and top soil. Warmer air holds more moisture and speeds up evaporation, which dries out the soil. Soil rich in clay solidifies when dry, making it hard for water to penetrate when it rains again, which leads to runoff, floods, and mudslides. Soil rich in humus becomes loose when dry, making it easy for wind to pick up the fines and carry them off in dust storms. Either way, whether by runoff, mudslides, or dust storms, topsoil is lost and agricultural productivity declines.

The power surge, the other side of the climate coin, manifests itself in extreme weather. Weather, over time, is an oscillation of rain and shine, heat and cold, insolation and precipitation. As greenhouse gases trap heat, they also capture energy from the Sun. Climate systems become more dynamic, which means that the amplitude widens and the oscillation magnifies. Wider amplitudes mean that peaks and troughs move farther apart vertically. Peaks crest taller and troughs bottom out deeper: when it rains it pours, and when it shines it burns. Magnified oscillations mean that the cycles stretch farther horizontally or in time: when it rains it keeps raining until it floods; when it shines it keeps shining until there is a drought.

Weather swings harder and hits farming with a double-whammy. In Siberia in 2010, the combination of global warming and power surge caused a heat wave joined by a drought. Fires broke out, and a third of the harvest was burnt. Russia stopped grain exports, which sent world food prices soaring. In winter 2010, poor nations experienced food riots. In spring 2011, arid countries dependent on imports suffered uprisings. Climate change turns our blue-green planet into a harsh world, with little water and less food.

The worst impact for life will likely be felt in the seas. Warmer waters are poorer in life. Cold waters are rich in plankton that sustains a high density of fish (which is why many fishing fleets operate in the high latitudes). Tropical seas have coral reefs teeming with fish, but such concentrations of life are the exception; farther out in the open and in the pelagic depth, low biotic density is the rule in warmer seas. Warm waters contain little plankton, and as temperatures rise, nutritive capacity falls further. The fish we harvest is the worst affected. Jelly fish is the least affected and has begun to displace commercial fish.

Oceans are carbon sinks. Since atmospheric $CO_2$ increases each year, more $CO_2$ now dissolves in the seas. This lowers the pH of water, making it acidic. Sour water attacks shells of calcifying organisms such as mussels and shrimp, the base of the marine food chain. As warming promises to turn oceans into the marine equivalent of a steppe, acidification may be an even greater risk, turning them into dead zones.

The short and the long of it is that the planet will stop feeding us. The humanitarian catastrophe in Darfur is a harbinger of things to come. Darfur's lands used to support farmers and herders as long as the monsoon kept fields and pastures green. Rising temperatures over the Indian Ocean destabilized this rhythm and make the monsoon erratic. Consequently, the land dried out; the age-old cooperation of sedentary and nomadic tribes exploded in genocide, and as early as 2007, UN Secretary-General Ban Ki-moon blamed climate change as one of the culprits of Darfur's collapse.[4] More failed states are to come.

## 2. Where do we stand?

Heading into the bottleneck and onto a new planet puts the question of *where do we stand* in perspective. The simple answer is: at a fork in the road. Either civilization keeps working with the dominant paradigm and enter failure mode, or humankind learns how to change – in a rapid, dramatic, and rational fashion.

Reaching the fork is a new reality of existence. And as current generations are the first humans in history who are faced with the fork, being-in-the-world has shifted. This relocation of existence relative to world is visible in the altered status of environmentalism, anthropocentrism, and climate. A closer look at these transformations sheds light on where we stand.

Consider environmentalism first. When environmental ethics joined the canon of disciplines 40 years ago, in 1971, with an inaugural conference at the University of Georgia, environmental protection appeared to the mainstream as an ideal, whose pursuit would be nice but costly. Now environmental protection is survival gear; it is the only way to keep doing business at all. As humankind moved ever closer to the fork, this protection morphed from an aesthetic luxury to an economic necessity and from a moral desideratum to a pragmatic strategy. Being at the fork means that stabilizing the Earth system, mitigating climate change, and adapting to the barren new world is the only way for us to avoid dieback.

At the same time, when environmental ethics was in its infancy, the question of who has moral standing was of prime importance. Environmental ethics was first and foremost about rights, especially about the rights of the nonhuman others. Specialists concerned themselves with the moral status of animals (and plants such as trees), and with the related task of adjudicating interspecies conflicts of interests. In those days, it seemed that environmentalism was an issue of the human 'us' versus the animal 'them'; that what is in our interest tends to undermine theirs, and what is good for our well-being tends to be bad for them. Now the 'us-versus-them' issue is moot. As our species is overwhelming, the capacities of the biosphere, and as environmentalism becomes survival gear, humans find themselves sitting in one and the same boat with nonhuman life. If the boat capsizes, we all go down. When they sink, we sink, and vice versa. Before the shift, anthropocentrism divided the concerns of classical ethics from those of environmental ethics. After the shift, at the fork, anthropocentric and nonanthropocentric concerns are converging.

Finally, the shift reverses the relation of climate change and environmental issues. Initially, climate change appeared as just another environmental problem. The earliest items on the list of environmental issues had been pollution, waste, and endangered species. Since 1971, the list has lengthened. More recent entries, in rough order of appearance, were nuclear power, acid rain, ozone layer, biodiversity loss, and resource depletion. At first, with papers such as Landsberg (1970) and Hansen et al. (1981), it seemed as if climate change would be yet another addition to the list, as an intangible and abstractly worrisome new thing. In retrospect, after the shift, we know it was not such an addition, not the least because climate change is not a 'thing'. The other list-items all fit in this ontological category. They are concrete and well-circumscribed phenomena (such as acid rain), technological risks (such as those associated with nuclear power), or linear dimensions of environmental decline, such as loss in biological diversity.

Climate change is categorically distinct from all of this because it is holistic, multifaceted, and integrative. This becomes evident as soon as we examine the concept. 'Climate' is nothing specific or concrete. It is not like weather. Instead, it is something indefinite and abstract; it is *the average of local weathers over time and across a region*. In terms of causation, it is the classical example of an ontological whole. Wholes are more than sums of parts because they are *the way* parts work together within a sum. A whole is to a sum what a dynamic outcome is to a static aggregate. The aggregate, here, is the Earth system, and its parts are five planetary domains, the regions of air, life, water, earth, and ice. The dynamic outcome of these parts set to work is climate: it is how atmosphere, biosphere, hydrosphere, lithosphere, and cryosphere interact. In this sense, it is a whole.

Accordingly, *climate change* cannot properly be called a phenomenon because its sweeping nature defies such a label. It is more appropriate to call it an emerging reality. This new reality, furthermore, has multiple facets in space and in time. Its spatial facets are, for example, changes of the arctic, the oceans, and the equatorial rain forest belt. Its temporal facets concern runaway cascades such as thawing permafrost soils, methane outgassing, seabed clathrate destabilization, and Antarctic ice sheet rupture.

Through its multifaceted character, the new reality constitutes a *risk spectrum.* For biomes, climatic perils involve ecosystems collapse and geophysical transformation. For species, such consequent dangers are attrition, displacement, migration, and extinction. For culture or civilization, risks arise in public health, ranging from intensifying allergies to the spread of tropical diseases; they arise in politics, ranging from food riots to uprisings to resource wars; they arise in economics, ranging from lost harvests to climate refuges to failed states; and overall, civilization risks range from social hardship to collapse.

The new reality of climate change informs virtually all phenomena on the list of environmental problems, plus spawning entire new orders of hitherto unknown troubles of its own. From the traditional vantage point of environmental ethics, it also affects whoever has moral standing in some form, whether these are people, future generations, apes, animals, plants, biotic systems, or Aldo Leopold's integrity of the land. Climate change, through its diverse facets, manifold risks, and multiple dimensions, is an integrative reality. It puts all the traditional problems in a new place. It arises as the salient context for all of them. Thus, it is not an entry on the list; it is the new paper the old items are written on. To put it baldly, *it is the list.*

Because being-in-the-world has arrived at the fork, everything is now different. As an academic aside, it is perhaps worth noting that environmental ethics is now obsolete. The sum-total of its subject-matter currently integrates in the existential context of climate change. Thus, *climate ethics* is its rightful heir.

From a philosophical look at the fork, all empirical trends point to the same conceptual conclusion: taking the right path – the path of sustainability, mitigation, and resilience – requires civilization to put as much distance as possible between itself and the paradigm whose implementation unleashed the climate crisis. The question of where we stand at the fork is also a question about location relative to the paradigm.

This paradigm is a bundle of cognitive modes that inform a view of nature in utilitarian and neoliberal ways. These modes originate in the cultural geographies of the Middle East and the Far West. Middle Eastern roots of the paradigm are the Abrahamitic or Judeo-Christian-Islamic belief-systems. Far Western roots are Scottish, English, and French variants of the Enlightenment. The Abrahamitic religions divorced divinity from nature, which drained nature of sanctity, and elevated humans above the rest of the creation, making them feel entitled to use the world as if it was their larder.

The Far Western Enlightenment, the Age of Reason, appropriated secular versions of this anthropocentric perception of nature. New strands added to the cognitive bundle, during this age, were a turn toward skepticism and an emphasis on individualism, which spawned liberalism. For fairness's sake, East European variants of the Enlightenment, such as the German *Aufklärung*, are mostly exempt from this blame, since their views of nature were more informed by holism and pantheism, often enhanced by a communitarian outlook and a turn to metaphysics. Thus, the modes of cognition that paved the way to the climate crisis, and which are in retrospect responsible for it, are prefigured by the likes of Moses and Plato, by Descartes and Voltaire, and by Smith, Locke, and Hume. The cultural set of their paradigmatic perspectives is the invidious triad of dualism, skepticism, and liberalism. Standing at the fork means civilization must make a decision about these figures and the perspectives they represent.

## 3. What does it mean to be human?

Knowing where we are heading and where we stand sheds light on *what does it mean to be human.* The current position of humankind is untenable. Doing business as usual is a recipe for failure. At the same time, we cannot turn back the clock. Hence, the only option left is *to evolve*. The primary evolutionary goal is obvious: we wish to ensure collective well-being. Subordinate to this end are three strategic goals that are equally evident, at least in principle: one, we must *adapt* to the consequences of climate change while trying to make them more bearable; two, we must *soften* or mitigate the unfolding of climate change while trying to head off causal cascades; and three, we must *rein in* and eliminate the causes of climate change while trying to sustain social and economic stability. The way to get there is through an emancipation of civilization from fossil fuels, economic growth, and unregulated markets. Evolution is accordingly a post-carbon, post-consumerist, and post-capitalistic transformation. Hence, to be human means to *mature*.

The carbon-based consumer culture of predatory capitalism, which is now becoming unsustainable, is a culture of immature development. The invisible hand of Adam Smith, the lazy skepticism of David Hume, and the unbelievable hubris of a Moses who told his people to be fruitful and multiply, and to fill the Earth and subdue it – these are the cultural equivalents of a raging teenager with attention-deficit disorder. Trust in the self-regulation of market forces discloses an Ayn Rand-like arrogance that we can rock it out and get away with it, and that acting as a selfish brat is the best thing that ever happened to the family. Denial of biospherical limits and of the etiology of climate change reveals a Humean cynicism that there are no universal links between cause and effect; that we do not need to worry about a thing as long as we cannot see it; and that warnings are for wimps. Such cynicism fuels epic fails and is typical of teenagers hell-bent on earning Darwin awards. The faith, finally, that the world is our oyster, and that we are following a manifest destiny, and that we are okay, because in God we trust, is the teenage attitude best illustrated by the most radically Christian country on the planet, USA.

Not that long ago, the American way of life was considered to be the high point of human development, and of such perfect maturation, for some particularly inebriated souls, that it spelled the 'end of history'. But now it turns out that this lifestyle is what unleashed the climate crisis in the first place. USA has the fattest *per capita* carbon footprint of any country on the planet (with the exception of tiny sovereign urban islands and a few oil-refining Middle Eastern sheikdoms). There is no other society worldwide with such excessive consumption, waste, and greed for energy. Although the US American population constitutes less than five percent of the world population, it has produced more than a third of the cumulative greenhouse gas emissions of our species. Anthropogenic climate change is really *Amerigenic* climate change. And despite the fact that USA perpetrated the global climate crisis nearly single-handedly, US policy has wallowed in denial. From Rio to Kyoto, from Bonn to Johannesburg, and from Bali to Copenhagen, the USA has acted like a crazed teenager strung out on oil, blocking or deflecting all efforts to adopt sensible and internationally binding emission caps.

Climate change asks civilization to grow up. The deeper answer to what it is to be human points to an obligation, in enlightened collective self-interest, of civil evolution. This illuminates the three questions stated in the beginning. Attending to this obligation means that we may *hope* to prevail as a civilization. It means we can *know* what is at stake. And it means we should *do* what it takes to get us there.

## 4. Plan B or climate ethics 2.0

This topic issue of *Journal of Global Ethics* is based on the assumption that conventional modes of thought are bankrupt. It is difficult to believe that the normative perspectives and conceptual tools that contributed to the climate crisis in the first place, such as discounting against the future, cost-benefit analysis, and other utilitarian market devices, will be capable of providing us with a solution.

Put differently, I doubt that 'Plan A' is going to work. A continued deliberation of market tools and utilitarian devices may not lead very far. What is worse, it appears that the prevailing mentalities of philosophical convention are not going to be of great use either. Philosophy, in the past century, split into postmodern thought and analytic philosophy. These two camps, despite their obvious differences, share the common ground of skepticism. Skepticism is inappropriate and irresponsible in the face of absolute limits. Postmodernity emphasizes culture over nature, deconstruction over structure, and interpretation over lawfulness. Such emphasis, it appears, makes things worse, as postmodern approaches are easily appropriated by climate deniers for corporate gain. Analyticity stresses the breakdown of information, the isolation of data, and the separation of events from context. This appears wrong-headed and obsolete, because climate change asks philosophers to do the precise opposite: the reconciliation of information, the understanding of data, and the integration of events in context. Analysis is not needed now; synthesis is.

The chapters in this volume explore philosophical alternatives to the dominant paradigm. All of the contributions are based on either 'thinking outside the box' or on 'critiquing its contents'. The rst chapter, by Michael Thompson, is an inquiry into the cultural maladaptation of American being-in-the-world and an argument of how Kantian imagination would point to civil evolution. The second contribution, by Byron Williston, is an assessment of the Canadian failure to rise to the occasion and an argument for moving toward virtue ethics. The third paper, by Ernesto Hernández, is an examination of Latin American thought in all of its current ambiguities – as a victim of the dominant paradigm and as a platform for the outlook needed to fathom the crisis. The fourth essay, by Bruce Janz, is an inquiry into the normative potential of Japanese thought for a global climate ethics. The focus of inquiry is the work of the rst philosopher of climate, Tetsuro Watsuji (1886 — 1960). The fth contribution, by Shih-yu Kuo, is an investigation of the heuristic power of Neo-Confucianism as regards the climate crisis and an argument for reappraising recent Western advances in environmental ethics as unwitting reformulations of far older Confucian positions. The sixth article, by Chen Xia and me, is an exploration of the promise of Daoist philosophy for engaging with the new planetary reality. Like the other essays, this article serves as an implicit reminder that the climate crisis asks us to escape the con nes of the Socratic love of wisdom, with its Anglo-

Greek proclivities of irony, doubt, and dissolution; luxuries we can no longer afford. The seventh contribution, by Jan Deckers, returns to the problem at hand, the self-indulgent and narcissist lifestyle in carbon culprit nations, especially in the USA. Using new empirical measures and universal moral considerations, Deckers analyzes the link between climate change and meat-eating, and shows how moving beyond a carnivorous diet would make an instant difference in mitigating global warming. The eighth chapter, by Tim Leduc, is an account of Inuit thought as regards arctic warming. The wisdom of a way of life attuned to cognitive limits in an interconnected world, which is encapsulated in the Inuit concept *silatuniq*, may shed some light on the post-consumerist path of progress. The ninth and nal chapter, by Sirkku Hellsten, Frederick Ochieng'-Odhiambo, and me, is an exploration of African sagacity as an alternative to unsustainable modes of thinking, such as economic rationality and the capitalistic model of development. Plan B, in the guise of global climate ethics, is the chance for philosophy to reinvent itself, as the rational inquiry of being-at-the-limit, sustained by world wisdom and the vision of civil evolution.

## Notes

1. For the first three queries, cf. Kant (1781, 1787), *Kritik der reinen Vernunft*, A804–5/B832–3; for the fourth, cf. Kant (1800), *Logik*, AK 9:25; for the last two, cf. Heidegger (2004) (1939/40), *Aussprache über Ernst Jünger*, GA 90:226.
2. 'Between the years 1900 and 2000 world GDP at constant prices has increased about 19-fold, corresponding to an average annual rate of growth of 3 percent'; cf. International Monetary Fund (2000, 151).
3. Brown (2011); cf. esp. ibid., 'On the edge', 3–18.
4. UN Secretary-General Ban Ki-moon, A climate culprit in Darfur, 2007.

## References

Ban, K.-m. 2007. A climate culprit in Darfur. *Washington Post*, June 16. http://www.un.org/sg/articleFull.asp?TID=65&Type=Op-Ed (accessed March 10, 2011).

Brown, L.R. 1998. The future of growth. In *State of the world 1998: A Worldwatch Institute report on progress toward a sustainable society*, ed. L.R. Brown, 3–20. New York/London: Norton.

Brown, L.R. 2011. *World on the edge: How to prevent environmental and economic collapse*. New York/London: Norton.

Hansen, J., D. Johnson, A. Lacis, S. Lebedeff, P. Lee, D. Rind, and G. Russell. 1981. Climate impact of increasing atmospheric carbon dioxide. *Science* 213: 957–66.

Heidegger, M. 2004. *Aussprache ueber Ernst Juenger*. Vol. 90 of *Gesamtausgabe*, ed. P. Trawny, 211–66. Frankfurt: Klostermann.

International Monetary Fund. 2000. *World Economic Outlook*. http://www.imf.org/external/pubs/ft/weo/2000/01/pdf/chapter5.pdf (accessed March 10, 2011).

Kant, I. 1781. *Kritik der reinen Vernunft*. 1st edition. Vol. 4 of *Gesammelte Schriften*, ed. Akademie der Wissenschaften. Berlin: DeGruyter, 1911.

Kant, I. 1787. *Kritik der reinen Vernunft*. 2nd edition. Vol. 3 of *Gesammelte Schriften*, ed. Akademie der Wissenschaften. Berlin: DeGruyter, 1911.

Kant, I. 1800. *Logik*. Vol. 9 of *Gesammelte Schriften*, ed. Akademie der Wissenschaften. Berlin: DeGruyter, 1962.

Landsberg, H.E. 1970. Man-made climatic changes: Man's activities have altered the climate of urbanized areas and may affect global climate in the future. *Science* 170: 1265–74.

# Climate, imagination, Kant, and situational awareness

Michael Thompson

*University of Tampa, Tampa, FL 33606, USA*

The interstate highway system and environmental are seldom discussed conjointly in works on climate and sustainability programs. In this essay I employ a metaphor, likening the interstate system to environments, to illustrate a cognitive shortcoming, a failure of imagination, by the organisms found in both. I argue that several failures of the imagination combine to constitute a failure to be aware of the limitations of our situations and the parameters set by climatological considerations. However, by re-engaging with our environment through the use of imaginative exercises, we can become aware of our surrounding in terms of both geographic and organismic limitations of our environment. Furthermore, to develop these sustainable policies, we can employ Kantian ethical deliberations that engage our sense of imagination to become 'situationally aware' of our place in the climates we inhabit and of others members of our environment.

Carazan's Dream: In proportion as his riches increased, this wealthy miser had closed of his heart from compassion and love toward all others. Meantime, as the love of man grew cold in him, the diligence of his prayer and his religious observances increased. After this confession he goes on to recount the following: 'One evening, as by my lamp I drew up my accounts and calculated my profits, sleep over-powered me. In this state I saw the Angel of Death come over me like a whirlwind. He struck me before I could plead to be spared his terrible stroke. I was petrified, as I perceived that my destiny throughout eternity was cast, and that to all the good I had done nothing could be added, and from all the evil I had committed, not a thing could be taken away. I was led before the throne of him who dwells in the third heaven. The glory that flamed before me spoke to me thus: "Carazan, your service of God is rejected. You have closed your heart to the love of man and have clutched your treasures with an iron grip. You have lived only for yourself, and therefore you shall also live the future in eternity alone and removed from all communion with the whole of Creation". At this instant I was swept away by an unseen power, and driven through the shining edifice of Creation. I soon left countless worlds behind me. As I neared the outermost end of nature, I saw the shadows of the boundless void sink down into the abyss before me. A fearful kingdom of eternal silence, loneliness and darkness! Unutterable horror overtook me at this sight. I gradually lost sight of the last star, and finally the last glimmering ray of light was extinguished in outer darkness! The mortal terrors of despair increased with every moment, just as every moment increased my distance from the last inhabited world. I reflected with unbearable anguish that if ten thousand times a thousand years more should have carried me along beyond the bounds of the universe I would still always be looking ahead into the infinite abyss of darkness, without help or hope of any return—. In this bewilderment I thrust out my hands with such force toward the objects of reality that I awoke. And now I have been taught to esteem mankind; for in that terrifying solitude I would have preferred even the least of those whom in the pride of my fortune I had turned from my door to all the treasure of Galconda—'.[1]

This stunning, Dickensian reminiscence predates the industrial revolution and Ebenezer Scrooge by nearly 100 years, but its foreboding warning finds traction from its original telling through the innovations of the nineteenth century and into our own time. And while its admonition once served as a reminder for personal compassion and awareness, today it serves as counsel for personal actions and national, even global, policy. Because we have acted individually, nationally, and sometimes globally in the manner of Carazan, we are facing the terror not of not only personal isolation and solitude but the end of the planet as we know it. In his dream, Carazan finds himself at the 'outermost end of nature' and today we, ourselves, are beginning to find an 'end of nature' because of our inattention to environmental considerations and moral turpitude in enacting policy change.

The chief cause of anthropogenic, more specifically Amerigenic, climate change comes as no surprise. Ostensibly, the cause of climate change is $CO_2$ emissions, this fact is incontrovertible.[2] What is more difficult to ascertain is the attitude that permits people, Americans specifically, to keep emitting $CO_2$ at their current, excessive level. Simply stated, the cause is narcissism. As early as the eighteenth century, Alexis de Tocqueville noted the American tendency toward individualism and resource consumption. Cautiously admiring as de Tocqueville was at times, he presciently warned what may come of this rising global power – that 'individualism, at first, only saps the virtues of public life, but, in the long run, it attacks and destroys all others, and is at length absorbed in downright selfishness' (see de Tocqueville 2000). By 1976, Erich Fromm boldly describes the final elaboration of the tendency noted by de Tocqueville – Western civilizations' occupation with possessing material wealth under the guise of unlimited progress and desire satisfaction (see Fromm 2009, 1–2). In the last decade we have become inundated with reports, accounts, and diagnoses of the Western/American attitude toward consumption and its environmental impacts.[3] Quite recently, a diagnosis of American (and Western) narcissism is to be found in Martin Schönfeld's essay 'Amerigenic Climate Change: An Indictment of Normalcy' (see Schönfeld 2010a). In short, the American love affair with its flawed worldview, its unprecedented consumption and resource acquisition and the lifestyle that it affords amounts to narcissism. And yet, unlike the Narcissus of myth, we no longer have the benevolence of Gaia (or other gods for that matter) that will transform our self-admiration into a sustainable, natural object. Americans, and all industrialized peoples, cannot depend upon a gracious mother nature in order to save them from collapse, but must look at the symptoms of their narcissism to diagnose the problem(s) and find solution(s). By examining Western lifestyles, I would like to argue that the chief attitude that continues these unsustainable practices is the same attitude that condemns Carazan, that is, narcissism and self-importance fueled by a lack of awareness to their surroundings.

In order to explore the narcissism that characterizes the American lifestyle and its subsequent symptoms in this paper, I shall first present what I find to be an apt analogy – that of the interstate as ecosystem – one that epitomizes the mind-set and the cognitive problematic of the quintessential American attitude. Integral to this analysis is the concept of 'situational awareness' and its collapse and expansion through the powers of the imagination. The interstate demonstrates how Americans have become alienated from their environment through the use of technology and a truncated ability to envision themselves in their actual surroundings. This metaphor, I believe, corresponds with the overriding narcissism found in Western lifestyles and their unsustainability in light of limited resources, rising global temperatures and $CO_2$ emissions, a narcissism that is due to the very same lack of vision. And yet, rather than simply elaborate the problem(s) of American lifestyle(s), I propose to present a propaedeutic to a possible solution. In so much as the imagination, with all its creative powers, indicates a crisis in our cognitive attitude, it also presents the means by which to overcome the problem. In this instance, the imagination holds the power to enframe our awareness, and, as Heidegger points out (see Heidegger 1993), Hölderlin's words ring true yet again:

But where the danger is, grows

The saving power also.[4]

The solution is neither technical nor practical, but, rather, indicates a shift in cognitive processing that will prove instrumental in achieving situational awareness and provides the ground for introducing environmental policy according to global, moral standards. Such a transformation of our attitudes demands a re-imaging and new found use of the imagination as a cognitive process that enables us to realistically enframe our understanding of the world in which we live.

## 1. Interstate as ecosystem

The interstate highway system is rarely discussed in environmental literature. When it does make an appearance, the interstate is most often discussed in terms of its impact on land use, air pollution, the division of roaming grounds for megafauna, and its role in human interaction with ecosystems.[5] Generally, these discussions make little positive use of the interstate, aside from potential human prosperity. However, I would like to suggest the interstate is an ecosystem itself. The interstate highway system is comprised of regularized boundaries of interaction, organisms interacting with one another, sub-systems of support, and a hierarchy of individuals on the road. The pavement/asphalt/cement itself delimits the terrain wherein organisms interact; gas stations, exits, rest-stops, and travel centers all support the commerce of travel; motorcycles, cars, trucks, and tractor-trailers constitute the organisms of the interstate; and these organisms are in constant commerce with one another – passing, changing lanes, and oft times colliding.

Whenever one hops on the interstate, a few obvious features stand out in regard to the interactions and relations of the organisms of this ecosystem. The automobiles and drivers are generally comported to one another in terms of safety- the primary concern of all drivers is to arrive at their destinations in one piece. And yet, the comportment of individual drivers seems to betray this overriding concern. Too often people exceed the posted speed limits – limits that are designated as the maximum speed to ensure safety. All too common, drivers change lanes without looking, tailgate, pass/change lanes at unsafe distances and fail to acknowledge the common commerce of the interstate – safe travel to destinations. One startling safety regulation most people nearly categorically ignore is the suggested distance between vehicles- roughly 10 feet per 10 miles per hour.[6] The simple diagnosis of the unsafe driving conditions is that mentioned earlier in this paper, that is, narcissism. People act as if they are the only organism in the ecosystem and their travel takes priority over all others on the road. So enamored are we of our own destinations that we fail to recognize the inherent dangers involved when barreling down the road at high rates of speed. In effect, we have isolated ourselves from communion with other organisms, failing to realize the demands of the situation that is the interstate and to follow proper safety protocols.

It may be the case that we are too engrossed in our own activities to recognize other drivers or the ecosystem that is the interstate; it may be the case that our individual concerns take priority over the situation, but such a simple explanations fail to capture all the elements that have led to our unsafe practices. Simply, narcissism writ large remains too easy a diagnosis – there are dimensions other than self-importance that factor into current driving trends.

Modern technology has led many drivers to feel quite safe in their automobiles, and rightly so (we do want to feel that we are safe). Advanced improvements in safety features have lessened the chances of fatality in our modern automobiles (see US Department of Transportation, National Highway Traffic Safety Administration 2005). Furthermore, technology has provided us with unprecedented comfort in our transportation. Air-conditioning, heated/massage seating, stereos, personal music devices, phones, noise reduction technology, global positioning systems, and now even televisions have all added a sense of comfort and ease in vehicular movement.[7]

Due to the time spent in our automobiles and our desire for comfort, we have, more or less, re-created one of our safest and most comfortable environments, our living rooms, in our automobiles. We have been wildly successful at re-creating, or imagining we have re-created, our homes inside our cars. The hermetic environment all too easily permits us to ignore the fact that we are traveling in fragile carriages made of metal, plastic, and glass and that we are moving at high rates of speed – the fact remains that when driving on the interstate we are hurtling nearly 80 mph in carriages made of aluminum, glass, and plastic. Because of these conveniences, our situational awareness fails, both in terms of the ecosystem as well as the organisms in the environment.

Moreover, these external and technological accoutrements are accompanied by a shift in attitude towards different ecosystems. Unfortunately, by reproducing our home-comfort in our automobiles, we have lost the awareness that the interstate is not our living room, and that the situational demands of the interstate ecosystem are not the same as those of being at home. What is at issue in the conversion of our automobiles into our living rooms is our attunement to varying situations and the situational awareness needed to safely traverse the interstate system. Plainly, too often, we fail to realize that the interstate is not 'ours', but is, rather, a space we share with other organisms – that is, we fail to recognize our ecosystem and the limitations it places upon us. This failure of recognition, I believe, poses serious consequences.

The chief consequence is that stated above, that we have begun to drive as if we are the sole possessor of the road, and that even should we meet with other organisms, their interests are secondary to our own. Other drivers have become mere obstacles preventing us from achieving our intended goals in an expedient manner – we have grown to disrespect each other on the road. Such narcissism sponsors unsafe driving and has, inevitably, led to countless collisions. Furthermore, this failure to attend to others as equal organisms in the ecosystem also indicates another failure that influences how we drive. By imagining our automobiles are our living rooms and the myopic attitude such familiarity brings, another explanation for unsafe driving practices is a failure to imagine what might happen should a collision occur. Because of our fabricated feelings of safety and comfort, we have begun to lose the ability to imagine that an accident/fatality could occur to each of us. As we barrel down the highway situationally unaware, content in our sense of self-importance, complacent in our feelings of safety and comfort, having lost the practice of imagining we occupy a dangerous environment, we often find ourselves passing those unfortunate enough to have reaped the results of such an attitude. The phenomenon of 'rubber-necking' demonstrates the case in point. It may be the case that onlookers of accidents are concerned about the well-being of those involved, and such compassion and possible communion is laudable. But we are also afforded a glimpse and reminder of the ecosystem that is the interstate. When moving past an accident, we need not imaginatively project what might happen in case of an accident, but have ostensible evidence of its outcome. The lucky few are those that can imagine themselves involved rather than strangers, and subsequently become situationally aware of the interstate again. Yet, too often, the response is one of morbid curiosity accompanied by the catharsis that 'it is not me' – and a failure to employ the imaginative exercise and the consequent awareness. Often, once we have passed the scene, we return to our ordinary narcissistic practices. A lack of situational awareness, by the success of re-imaging our homes in our cars, has led to a distinct failure to imagine collisions can happen 'to me'.

If one understands the interstate as an ecosystem and comprehends the various elements at work that assist in continuing our narcissism, one realizes that by driving unsafely and erratically, ignoring the other organisms on the road and the demands of the environment, we are employing unsustainable practices – people are wrecking. This microcosmic analysis of human behavior can be extended to the macrocosmic scale – the same narcissistic attitude found on interstate highways is found in our attitudes and practices concerning the global environment. By failing to imagine/recognize we are not the only organisms on the planet, by failing to be situationally aware of global warming, we are wrecking the only world we inhabit.

## 2. Re-imagining our world

Much like our situational awareness, or lack thereof, on the interstate, the diagnosis and explanatory mechanisms of narcissism apply to the typical Western (quintesessentially American) attitude towards their environment and the current environmental debate. Few, if any (Monbiot, 2007, 20–42), researched, scholarly studies of current green-house gas emissions deny that humans are in fact adding to $CO_2$ emissions to the atmosphere and subsequently contributing to rising global temperatures. Anthropogenic, specifically Amerigenic, causes of global warming are incontrovertible. But what remains questionable is the attitude that permits an obvious fact from informing national – and even global – policy. Despite the overwhelming documentation of the role humans and Western lifestyles play in GHG emissions, what permits us to be oblivious to and/or ignore our situation remains a question that must be addressed. Narcissism, and a subsequent lack of situational awareness, is, I believe, the appropriate response to such a question.

American acquisition of resources and technology has heralded an unprecedented standard of living. The fact that heretofore sparsely populated regions of America now occupy a rising population, e.g. Florida, Nevada, Alaska, etc.[8] attest to humankind's ability to master their environment, while subduing (or at least keeping at bay) usually inhospitable climates. Much like our automobiles, technology has afforded the luxury of constant temperature regulation, much to our ignorance of the natural climes. Air-conditioning, both heating and cooling, are the main contributors to our mastery of seemingly inhospitable environs. In addition, much like the interstate, these technological innovations have added to our perceived notions of safety and comfort. For example, an Alaskan may bask in the luxury of 70 degree temperatures year round in the confines of their homes, while ignoring the constraints of their environment. Quite literally, populations in tundra or even arctic climate zones can enjoy temperate climates within their homes, even in the frigid months of winter, while imagining themselves in tropical or sub-tropical temperatures. Conversely, those in hot, tropical climates can effectively ignore their environmental conditions so long as they remain within their air-conditioned buildings and automobiles. People inhabiting places like Florida and Nevada can ignore the brutal heat of the summer cycle. That such technology is useful finds no dispute, but one must ask: 'At what cost?' Fuel shortages and escalating prices immediately impact consumers in these locales, but the cost is not merely economic. The deeper, more troubling and personal cost is, as on the interstate, an increased sense of self-importance, a lack of situational awareness and the attitude that one's personal needs outweigh any detrimental environmental impact.

Just as the example of the interstate has demonstrated, the role of the imagination in such narcissism remains ambiguous. The ability to re-create an artificial climate capable of sustaining life and affording desire satisfaction, while ignoring the actual environmental constraints, bespeaks a remarkable cognitive success for the imagination. For the most part, because of our technological advances, we now have the ability to inhabit inhospitable climes all the while ignoring that they are dangerous. And yet, at the same time, the myopic viewpoint that each individual's luxury takes precedence and needs be satisfied, marks a failure on two levels.

The initial failure of our cognitive imagination is to envision ourselves as a part of a larger ecosystem, one within which we are only a mere member and not the sole or single most important organism. Ostensibly, we do not require much imagination to understand our place within the order and hierarchy of our ecosystems. But due to the hermetic sealing and isolation that certain environs may elicit, or merely the desire for increased control of our environment, we often cut ourselves off from both other organisms and the ecosystem itself and fail to see ourselves a part of a larger environ. Even in the most frigid or hottest climates, one rarely shuts out the environment completely. If for no other reason, the fluctuations in our electricity bills remind us of the larger system of which we are a part and to recognize other entities that also participate in our environment. Thus one is nearly forced to periodically remember one's role and imagine where one fits within the larger whole. And yet, this is precisely the first failure we find in the narcissistic attitude. Because of our truncated existences, we often no longer recognize other organisms as members of the ecosystem, or the ecosystem itself. Electric companies are not peopled by other humans, they are just faceless entities; our surroundings are hostile and to be avoided at all cost – in both instances they are no longer images that elicit recognition and this is due to an over-inflated sense of self-importance.

The second failure of the imagination in this larger context mirrors the second failure found in the interstate example. In the microcosm of the interstate we failed to imagine the potential consequences of our unsafe driving practices, in the macrocosmic picture we fail to imagine the potential consequences of unmitigated resource consumption and emission of $CO_2$. In both instances the failure is the same; rather than recognizing we are practicing unsustainable habits, we eschew any responsibility for our unsafe practices. The chief manifestation of such an attitude is a purposive disavowal of consequences and a willful ignorance that our actions may have personal or global impact. The narcissistic attitude promotes a form of separatist thinking that enables us to deny or ignore that our own actions have any impact on others. Much like vehicular accidents, we hear/see reports of the impacts of rising global temperatures. And yet, just like in vehicular accidents, we merely note them with morbid curiosity and fail to recognize them as the eventuality of environmental constraints. Unlike the immediacy of automobile collisions, we find examples far removed from our comfort-created homes and the distance permits us to fail in being aware of global climatic changes. Seeing images of climatic changes – e.g. climate induced war in Darfur, starvation in third world countries due to desertification, even polar bears losing their habitats – provides a glimpse into the potential damages of unsafe policies, and yet exclusionary thinking and the comfort we enjoy prevent us from imagining that such consequences will affect us or that we may be in part responsible. Such failures are manifest in the dispositions that deny that global warming will have any impact (if there be any at all) on those of us living today. This 'not in my lifetime' attitude bespeaks another aspect of imaginative failure – that of failing to realize we are leaving future generations with a potentially insoluble problem. Such a failure to project into the future the consequences of our current actions is perhaps the most pernicious form of the narcissistic attitude. It betrays in the starkest light the common circumstances of failing to be situationally aware – both now and into the future. Such overt narcissism in effect denies recognition to anyone or anything other than our own immediate concerns. Moreover, this imaginative failure results in a lack of situational awareness not merely on a local level, but now on a global scale. Rather than realizing the rising global temperatures and its effects, we just turn the thermostat to a cooler setting, and in so doing we are wrecking our planet.

Such comfort and hermetic isolation contributes greatly to our egocentric attitude and produces a culture of narcissism. And, as easily seen, narcissism is most certainly a dangerous attitude to adopt, both in driving and in terms of environmental practices. But rather than merely condemn the Western lifestyle(s) and call attention to its pervasiveness in our lives, I believe the forgoing analysis can provide some hope. This hope springs from recognizing the decisive cognitive function that permits continued narcissism and a lack of situational awareness that obtains in the equivocal use of the imagination. Thus, in order to restore situational awareness, promote sustainable environmental (and traffic) practices and break the spell of technologically induced narcissism, recourse to this often overlooked capacity will be vital.

## 3. Imagining a solution

In order to disambiguate the functions of the imagination and its role in promoting narcissism (as opposed to a function for positive force), and thus to find redemption from our narcissism and global catastrophes, one must differentiate the reproductive role and the productive role the imagination plays in our cognitive processing. Fortunately, Immanuel Kant has established the precedent in his 1st critical work, *The critique of pure reason.* In his magnum opus, Kant distinguishes the reproductive mode of imagination from an original, fundamental power of productive imagination. In the former, one merely re-creates or reproduces events, images or concepts previously established by our cognitive processes; whereas, in the latter, one finds a truly creative process by which we organize our world and through which one can establish principles for action. Unfortunately, the scope of this essay does not permit any finer grain analysis. But, for the purposes of environmental policy and an escape from our narcissistic tendencies, it is to the latter's function and its relationship to ethics and sustainability I wish to turn our attention.

Kant's specifically ethical writings are chiefly found in his *Groundwork for the metaphysics of morals* and *The critique of practical reason.* In these two volumes, Kant establishes the priority of the moral law, a moral law that ignores biographical or situational exceptions, and, instead, emphasizes universality, respect for organisms (human organisms) and the goal of creating a world wherein all agents act and are treated as autonomous individuals acting within an ecosystem, with equal place of importance and consideration for all such individuals. In Kant's formulations, we find the moral law comprised in triplicate to express different facets of the same goal of morality and responsibility.

Kant's first formulation is found in his universal law formulation:

(1) Act only in accordance with that maxim through which you can at the same time will that is become a universal law.[9]

In itself this prescription calls for a break from individualistic thinking and summons our powers of imagination to project the possibility of global implementation of any action under deliberation. Because autobiographical and situational details must be removed in testing maxims under deliberation, personal and contextual details are removed, thus providing an initial opportunity to break with the ego-centric narcissism found in everyday practices. By following Kant's suggestion, one must envision what the consequences of universal implementation of considered policy might produce. Should implementation of the action under deliberation produce a paradox wherein the action under review is simultaneously made impossible by implementation, the action must be deemed morally impermissible.[10]

The first formulation of the categorical imperative provides an argument that all our actions must be sustainable – the true litmus test is not whether sufficient resources are adequate for the action, but, rather, whether the action itself, outside of material constraints, is a sustainable policy.[11] Failing this imaginative test, the action under deliberation must be deemed unethical and ought not to be implemented. This first formulation requires an initial programmatic of removing our biographical and contextual selves from the ethical decision making process – that is, imagining the consequences of our actions removed from our personal desires. We ought, rather, to view our actions in terms of conceptual-global feasibility, and this may provide a possible counter-movement to the sustained, hermetic isolation of narcissism. No longer shall we attend to the desires of our myopic self-interest, but now must imaginatively envision other organisms in our ecosystem acting in like manner. However, Kant's first formulation still tends to emphasize the individual and his/her cognitive processing, albeit now with a universal perspective, and such a methodology may tend to reinforce the isolation of individuals. Fortunately, Kant provides us with a second formulation of the categorical imperative, one wherein we find a supplement to the sustainability argument from the first, a supplement that will shatter the self-importance of narcissism by explicitly focusing on the other organisms in our ecosystem.

Kant's second formulation of the categorical imperative, the moral law, states:

> (2) Act so that you use humanity, as much as your own person as in the person of every other, always at the same time as end and never merely as means.[12]

This second prescription urges us to move beyond a merely theoretical construct of imagining universal implementation of our maxims (actions under deliberation) to an applied orientation of our interaction with others. In order to implement this second formulation, recourse to a veritative reproductive imagination is required, that we may now regain an awareness of the organisms with which we have interaction in our ecosystem(s). What such a prescription demands of us is that we no longer see other organisms as mere obstacles to our own designated goals/ends, but, rather, must now properly estimate the beings with whom we have interaction and treat them with the respect that other autonomous agents ought to be afforded. In our narcissistic attitude, a faithless imagination permits us to represent others as mere instruments/obstacles and to ignore other organisms as equal travelers through our ecosystem, while dismissing them as mere means (usually obstacles) to our own personal goals. Other drivers, for example, share the interstate with us because they too have destinations and goals in mind and traverse the roadways to effect the accomplishment of them. Becoming cognizant of other drivers, not as mere objects around whom we must navigate but as other subjects with whom we share our habitation, shatters the insulating imagining of self-importance and permits us to become situationally aware of the players within our dynamic environment. Kant's recommendation here is that we should continuously envision our world as peopled with other autonomous agents (as it actually is). Furthermore, by imagining not merely those immediately in our vicinity, but also throughout the entire world, we shall become cognizant of the possible global impacts of our local actions. Through this exercise, no longer can we dismiss the impacts of global warming and ignore its detriment on persons in foreign lands. At last we must face the truth that our local actions may impede others' lives (goals) and in so doing must be deemed unethical – at last we must become situationally aware of the limitations of our ecosystem, both in terms of natural constraints and other organisms.

Kant's third formulation of the categorical imperative provides us with the most expansive, and perhaps the most originally creative, employment of the imagination. According to Kant, by universalizing our intended actions and treating other agents as autonomous, we shall obtain

> (3) a very fruitful concept depending on it, namely that of a realm of ends ... the systematic combination of various rational agents through communal laws.[13]

This last formulation of the categorical imperative, while not possessing the prescriptive force of the first two, recommends that one ought to think and act in such a way as to bring about a kingdom of ends. Such a realm is one wherein all rational agents are availed the liberty to act as universally legislative entities, and by doing so should perform their ethical deliberations according to the first two formulations of the categorical imperative. By providing the opportunity for autonomous, rational agents to perform their own moral deliberations, all the while concluding with similar judgments, every deliberator is respected equally and arrives at the same sustainable, ethical policies. But arriving at such a kingdom of ends requires an almost superhuman exercise of imagination to break free of our narcissism. Indeed, it may be difficult to envision other organisms as rational or as autonomous (having their own specific ends – even if they be merely natural functions). But, in permitting others the ability to deliberate for themselves, we now create a world wherein our situational awareness is accompanied by the proper response to such organisms. Moreover, in a second case, the case of non-rational beings, we find a newfound respect for nature and the organisms of our global ecosystem (as they too have ends designated by their natural functions). By considering both rational and non-rational beings as belonging to the realm of ends, one breaks the shackles of narcissistic thinking and the vestiges of both self-importance and the often consequent paternalism falls by the wayside. In effect, by pursuing a world wherein we respect both other agents as well as natural organisms, we imagine a world in which both humans and other organisms are now considered a part of our ethical topography. By becoming aware of both species of organisms, that is by becoming situationally aware of our environment and its inhabitants; by envisioning a world in which all are treated with respect; and by pursuing a kingdom of ends, we will transform our own attitude and find an non-narcissistic, ethical ground that will encourage sustainability. Such an attitude endorses a truly global legislative process, wherein all constituents come under consideration.

## 4. Conclusion

The fact that humans impact their environment and each other accompanies the fact of our existence. With this existence comes a myriad of orientations for both our interpersonal and environmental interactions. At our current point in the development of Western civilization, we find ourselves at a decisive moment. Human technology and desire has provided unprecedented safety and luxury, but at the expense of environment. $CO_2$ and other green-house gas emissions and dependence of $CO_2$ emitting energy sources have produced the conditions for anthropogenic climate change. Already, we have begun to see alterations in our atmosphere caused by our life-style(s). At our current level of $CO_2$ emissions, cataclysmic environmental changes will become evident all too soon. The causes of these changes and the eminent signs of danger come as no surprise. What is surprising, however, is the inattention such signs elicit. Nominal action has been taken to 'prevent' further emissions and politicians emit much gas in discussing possible policy change. What is conspicuously lacking, however, is responsibility at local and personal levels, the levels at which the most effective change can take place. We have become Carazan-like in our attitude towards these issues – we have 'closed [our] heart[s] to the love of man and have clutched [our] treasures with an iron grip. [We] have lived only for [ourselves], and therefore [we] shall also live the future in eternity alone and removed from all communion with the whole of Creation'. Yet an even greater indictment is implied, our truncated lifestyle has created the condition for the end of the world as we know it. The American attitude is one in which, if not outright denying the role of $CO_2$ emissions on atmospheric conditions, people find themselves uncaring and unwilling to alter their own local habits in favor of global concerns. The root of this apathy is easily diagnosed as a narcissistic attitude that pervades Western culture and permits only immediate ego-centric concerns. Thus it is not merely $CO_2$ emissions that are causing environmental change, but, more importantly, the attitude that permits us to continue our unsustainable levels of emission. Carazan's narcissism is our narcissism and we may suffer the prescribed fate.

Such narcissism is, in its own right, the consequence of myopic thinking and an inability/unwillingness to become attuned to our environment. And yet, it is not merely the inability/unwillingness to become attuned to our global environment that remains the problem. The trenchant attitude of narcissism abides at even the local scale. Interstate commerce, electricity usage, consumption of resources and many others ... all of these instances bespeak a fundamental failure on our part to be aware of the circumstances in which we find ourselves. Such myopia can easily be blamed on technology, but technology itself has no intentional states. The blame must firmly be placed upon the shoulders of those who employ technology in ways that permit them to become ignorant of their environment, that is, to lose our situational awareness. Inherent in our cognitive processes is the ability and desire to imagine ourselves in places unlike those in which we actually find ourselves, and technology has enabled us to actualize these fantasies. Such is the difficulty and diagnosis of much of our narcissistic tendencies – we imagine we are the single most important consideration and even go so far as to re-imagine/re-configure our actual surroundings more to our liking. Such an attitude is unrealistic. The consequence of the various failures and successes of our imagination(s) is to become closed off from our situation/environment/ecosystem(s). And yet, precisely where we fail, we may find hope for the future. By becoming aware of our own powers of imagination; by becoming aware of our ordinary, truncated existence(s); by becoming aware of our own lack of situational awareness – we may find the way by which we can develop an environmental ethic. In order to do so, we must shed our historical development of individualism and its concomitant selfishness, that we may re-envision ourselves in our ecosystems and adopt truly sustainable practices. Since we may no longer appeal to a benevolent mother nature to save us from our own attitudes, we must effect a transformation from within. Such transformation is aided by accurately envisioning ourselves in the world, by re-cognizing the organisms in our environment and by imagining a world wherein these organisms pursue their own ends. I hope that in the foregoing analysis of situational awareness and its difficulties, we may find a new opportunity that loosens us from our American attitude, from an anthropocentric narcissism, and may provide the means to establish an ethical orientation that promotes equality and respects both humans and nature along Kantian lines.

## Notes

1. From Kant's (1960) *Observations on the feeling of the beautiful and sublime*, 48–9 fn.
2. See Intergovernmental Panel on Climate Change (2001, 4–5). Also, Tim Barnet, quoted in Pearce (2005).
3. To name but a few Monbiot's (2007) *Heat* and Garvey's (2008) *The ethics of climate change*.
4. From "Patmos" (1966) cf. *Friedrich Hölderlin poems and fragments*.
5. For example, Waddell's (2002) Modeling urban development for land use, transportation, and environmental planning; Mader's (1984) Animal habitat isolation by roads and agricultural fields.
6. Despite the seemingly great distance, even this is a conservative estimate – if one observes 'the three-second rule', the minimal safe distance between vehicles driving 80 mph is 352 ft, 60 mph is 176 ft. According to the American Association of State Highway and Transportation Officials and the US Department of Transportation's Federal Highway Administration, the average person needs a minimum of 2.5 seconds in order to react and respond to road conditions cf. p. 3 2001 *Green Book.*
7. Allan Bloom documented this tendency over two decades ago in his *Closing of the American mind* (1987). While Bloom's analysis pertains to portable music devices, the result in the same, a cutting off from and ignorance of our surrounding and its inhabitants.
8. Census 2000. Nevada, Arizona, Florida, Alaska ranked #1, 2, 7, 16, respectively, in the nation for population growth 1990–2000.
9. Kant (2002, Ak. 4:421); also *Critique of practical reason* 4:31.
10. Kant (2002, Ak. 4:403); also *Critique of practical reason* 4:27.
11. For more on the categorical imperative and sustainability cf. Schönfeld's (2010b) "Metaphysics of sustainability."
12. Kant (2002, Ak. 4:429).
13. Ibid. Ak. 4:433.

## References

Bloom, A. 1987. *The closing of the American mind.* New York: Simon & Schuster.

de Tocqueville, A. 2000. *Democracy in America.* Bk II Ch. 1, pp. 120–1, Bk II Ch. 9, pp. 150–1. Trans. and ed. H.C. Mansfield and D . Winthrop. Chicago: University of Chicago Press.

Fromm, E. 2009. *To have and to be*. London: Continuum.

Garvey, J. 2008. *The ethics of climate change: Right and wrong in a warming world.* London: Continuum.

Heidegger, M. 1993. Question concerning technology. In *Basic writings*, ed. D.F. Krell, 311–41. New York: HarperCollins.

Intergovernmental Panel on Climate Change. 2001. *Summary for policymakers to climate change 2001: A synthesis report of the IPCC Third Assessment Report*. London: IPCC.

Kant, I. 1960. *Observations on the feeling of the beautiful and sublime*. Trans. J.T. Goldthwait. Berkeley: University of California Press.

Kant, I. 2002. *Groundwork for the metaphysics of morals*. Trans. A.W. Wood. New Haven: Yale University Press.

Mader, H.J. 1984. Animal habitat isolation by roads and agricultural fields. *Biological Conservation* 29: 81–96.

Monbiot, G. 2007. *Heat: How to stop the planet from burning*. Cambridge, MA: South End Press.

Patmos. 1966. *Friedrich Hölderlin poems and fragments*, trans. M. Hamburger, 462–3. Ann Arbor: University of Michigan Press.

Pearce, F. 2005, June 11. Climate evidence finds us guilty as charged. *New Scientist* 186, no. 2503: 17.

Schönfeld, M. 2010a. Amerigenic climate change: An indictment of normalcy. *Human Ecology Review* 17: 117–24.

Schönfeld, M. 2010b. Metaphysics of sustainability: Kant's categorical imperative. In *Sustainability and the quality of life*, ed. C. Tandy and J. Lee, 1–18. Palo Alto, CA: Ria University Press.

US Department of Transportation, National Highway Traffic Safety Administration. 2005. *Trend and pattern analysis of highway crash fatality by month and day* (Technical Report, DOT HS 809 855). Washington, DC: National Center for Statistics and Analysis. http://www-nrd.nhtsa.dot.gov/Pubs/809855.pdf.

Waddell, P. 2002. Modeling urban development for land use, transportation, and environmental planning. *Journal of the American Planning Association* 68: 297–314.

# Moral progress and Canada's climate failure

Byron Williston

*Department of Philosophy, Wilfrid Laurier University, Waterloo, ON N2L 3C5, Canada*

In a recent letter to Canada's national newspaper, *The Globe and Mail*, British columnist and climate change gadfly George Monbiot pleaded with Canada to clean up its greenhouse gas emissions act. The letter appeared just a week before the Copenhagen climate conference. In it, Monbiot alleged that Canada's newly acquired status as oil superpower threatens to 'brutalize' the country, as it has other oil-rich countries (Monbiot, G. 2009. Please, Canada, clean up your act, *The Globe and Mail*, November 30, A15). In this paper, I want to expand on Monbiot's bleak assessment of the Canadian national psyche. It has been pointed out that climate change is forcing us to rethink philosophical ethics. Some, like Dale Jamieson, believe that virtue theory is best equipped to meet the challenge of understanding the moral dimensions of this phenomenon. I think this is basically right, but that climate change is also forcing us to reassess our capacity for moral progress. The two challenges are linked. In what follows, I will first (Section 1) motivate the appeal to virtue ethics as a new way of understanding the ethics of climate change. Next (Section 2), I offer a virtue ethical account of moral progress. With the latter in place, we can (Section 3) uncover the real nature of Canada's moral failing on climate change: it is an impediment to the moral progress of our species.

## 1. The abrupt paradigm and the turn to virtue ethics

According to many scientists, there are reasons to believe that some of the predictions of the 2007 report of the Intergovernmental Panel on Climate Change (IPCC) are too conservative. Recent climatological and meteorological data suggest that climate change is progressing much more rapidly than the IPCC thought it would. One general reason for this is that the IPCC did not take full account of the effects of various positive feedbacks in the climate system. This is how the recent Copenhagen Diagnosis puts the point: 'recent observations show that greenhouse gas emissions are changing near the upper boundary of the range of IPCC projections'.[1]

The two most alarming indicators of this new reality are probably the extent of Arctic sea ice and global ocean temperatures (though there are plenty more problems). The first concern is focused primarily on the amount of Arctic Ocean area covered by ice in summertime. Between 1980 and 2007, this area went from 10 to 4 million $km^2$. At this pace, the Arctic Ocean will be ice-free in the summer within 15 years. The IPCC predicted that this would not happen before 2050. No more ice up there means that the bulk of sunlight falling on that vast ocean will be absorbed by it, rather than reflected back to space. This will accelerate the heating process dramatically.

The second concern has to do with global ocean temperatures. There are new satellite observations of ocean areas that show a dramatic decline in the population of ocean algae. The dead area comes up a deep blue in the photos, rather than a blue-green. This area has increased in size by 15% over the last 9 years. The reason is that the surface of the ocean is warmer than it was previously, which means that the nutrient-rich cooler water from below has a harder time mixing with it. So it becomes more difficult to support life there. The growth of algae in the oceans cools the planet by drawing down carbon dioxide. Less algae means more carbon stays in the atmosphere (Lovelock 2009, 27–9). Again, the heating process is accelerated, but the IPCC did not consider the impact of losing this particular carbon sink.

Observations like these have not escaped the notice of philosophers writing on this issue. Some have been led by them to abandon the 'gradualist paradigm' of climate change and embrace instead the 'abrupt paradigm' (see, e.g. Gardiner 2009; McKinnon 2009). According to the former, the rate of climate change correlates relatively smoothly with changes in atmospheric concentration of greenhouse gases. It is a linear process of change which, as such, is relatively easy to adapt to. According to the abrupt paradigm, however, there are climate tipping points which, if triggered, could lead to events that will boost global average temperatures to catastrophic levels. James Hansen provides a useful distinction to capture the peculiar peril of our situation. We are, he says, between a 'tipping point' and a 'point of no return'. Our having reached the climate tipping point means that greenhouse gas concentrations are high enough to make it likely that there will be climate-caused catastrophes. This is what justifies the switch to the abrupt paradigm. The point of no return, however, would have been reached when we have lost all hope of reversing these processes (Hansen et al. 2008). This will increase the frequency and severity of climatic catastrophes and put us fully into adaptation mode.

So the abrupt paradigm is a reality, and the key to avoiding the slide to a point of no return is to cleave to the 2°C guardrail. This, however, will not be easy:

> If emissions were to continue to grow past 2020 ... holding the 2°C line would ultimately require extraordinary rates of emission reduction, rates corresponding to such large and historically unprecedented rates of accelerated capital-stock turnover that, frankly, it's difficult to imagine them occurring by virtue of any normal, orderly economic process. Time, in other words, is running out. (Kartha et al. 2009, quoted in Gardiner 2009, 154)

And yet in the face of the looming catastrophe, we are doing very little. We are stuck in a state of deep political inertia, exemplified most clearly perhaps by the Copenhagen debacle in December 2009. This inertia is, of course, merely a symptom of a deeper moral failing, a kind of 'moral corruption' as Stephen Gardiner would have it (Gardiner 2009, 140). What is the source of this corruption? For Gardiner, it is a collective action problem exacerbated by the fact that the chief objects of potential moral concern, i.e. future generations, do not yet exist, and so have no bargaining power with us. Nor does the abrupt paradigm solve this problem, wedded as it is to the notion that some degree of catastrophe might be unavoidable. To the contrary: Gardiner argues that to the extent we believe this we will relax efforts at mitigation – activity aimed at improving the lot of future generations – and focus exclusively on adaptation. Moreover, even though some adaptations will benefit future generations, most will likely be the sort of quick-fixes that are meant to secure only the interests of the present. And each successive generation will behave this way. Each will burn as much fossil fuel as it deems necessary to cope with its own problems, thus exacerbating climatic conditions for every future generation. This is the essence of what Gardiner calls the 'intergenerational arms race' (Gardiner 2009, 150–3).

So if we are in the midst of a vast collective action problem, we need a way of thinking that can break us out of it. The signal feature of collective action problems is that the participants employ a form of calculative thinking that nearly always produces suboptimal outcomes. Every agent's action is contingent on beliefs about what other agents might do. The chief goal is to improve one's comparative advantage or at least to avoid slipping in this regard.[2] The herdsman on the commons thinks obsessively about what the other herdsmen will do with all that lovely pasture; the quarantined prisoner about what the other prisoner is pondering. This way of thinking makes fear of being a sucker and desire to be an exception our guiding practical drives. These are demonstrably faulty cognitive structures and they need to be set aside for a problem like climate change. In their stead, we should decide what sorts of people we want to be and behave accordingly. In short, we need to develop a virtue ethical standpoint for climate change. Here is how Jamieson puts the point:

> Instead of looking to moral mathematics for practical solutions to large-scale collective action problems, we should focus instead on non-calculative generators of behaviour: character traits, dispositions, emotions . . . When faced with global environmental change, our general policy should be to try to reduce our contribution *regardless of the behaviour of others*, and we are more likely to succeed in doing this by developing and inculcating the right virtues than by improving our calculative abilities. (Jamieson 2007, 167, my emphasis)[3]

Note that as a practical strategy, this does not require us to forgo *bargaining* with others, something a superficial reading of Jamieson's advice might suggest. For instance, there is no reason to think that we could not be virtue ethicists and enter into emissions reduction negotiations with others in an effort to coordinate our proposed actions in the most efficient manner possible. Such negotiations might be as tough and substantive as you like. It is just that our acting in morally justifiable ways would not be contingent on whether or not others were willing to do so. A person with the appropriate set of green virtues would simply not countenance certain proposals, but otherwise there is room for plenty of flexibility about how best to resolve what remains, after all, a problem of collective action.

To be sure, part of the point of turning to a stable set of virtues or dispositions to anchor right action is precisely to ensure that agents are not flexible all the way down, as it were. For Jamieson, this means that an element of moral complacency is likely to mark the virtuous agent. But I would say instead that the virtuous agent will not be distracted by fear and cupidity and, more importantly, that this fact about her will be known by relevant others. This will alter the way they negotiate with her. They will be forced to couch their appeals and proposals in terms amenable to the agent's perceived sense of what is just or compassionate or temperate. With the focus on comparative advantage gone or dramatically diminished, there is more opportunity for morally justifiable policies to emerge from such negotiations.

The move to virtue ethics on this issue has two further advantages, not noticed by its advocates. First, and most obviously, virtue ethics provides a fruitful way of thinking about climate change under the abrupt paradigm. This is because the abrupt paradigm is an intensification of the problems that define the gradualist paradigm. Again, on the gradualist paradigm, it is difficult to solve the problems posed by climate change chiefly because each country is trying to maximize its own short-term interest in a manner that is relatively unconstrained by consideration of the interests of future generations. The same is true under the abrupt paradigm but now with the stakes ramped up significantly.

That is, to the extent the abrupt paradigm indicates that some climate-related catastrophe is likely on the horizon, it will foster a kind of siege mentality among the relevant 'players'. This mentality, in turn, will cause countries and their citizens to scramble for whatever means they can find to cope with their own perceived climate-related challenges. And all this activity will be keyed to, contingent on, beliefs about what other relevant players are doing. As we have seen, consideration of the interests of future generations will likely be an even less potent moral constraint in this scenario than on the gradualist one. But if the sort of calculation Jamieson is talking about is the most salient impediment to right action and policy under the gradualist paradigm, then this is so *a fortiori* under the abrupt paradigm. It follows that if the turn to virtue ethics is sufficiently motivated in the former case, it is an even more pressing desideratum in the latter.

Second, climate change is clearly forcing us to think very seriously about our own prospects for moral progress. There is undoubtedly some reason to believe that humanity has made moral progress in the last 100 years or so. Those who think this way tend to point to achievements like the end of slavery, the emancipation of women in some parts of the world, the broad and increasing reach of human rights discourse, the practices associated with it and so on. The most natural way to think of these achievements is that our moral vision underwent some *expansion* when we discovered that some subset of humanity – or some other entity altogether – had moral standing. This usually works by way of a realization that what we took to be morally relevant properties – race, class, gender, the capacity for language – did not in fact mark a barrier between those with moral standing and those without. It is possible to celebrate these expansions without being either triumphalist about the past or Pollyannish about the future.

Climate change clearly brings us face to face with a new group of moral claimants: future generations.[4] We are behaving as though their current non-existence is a property which deprives them of moral standing. However, given our power to alter their world dramatically, we will not take the next step in our moral progress until we disabuse ourselves of this thought and the immoral practices that go with it. So if we need to turn to virtue ethics to think our way out of the impasses brought on by climate change, and if climate change is testing our capacity to progress morally, the question is whether or not virtue ethics can offer a plausible account of moral progress. I think it can.

## 2. Moral progress and the expansionist ideal

Richard Rorty has argued that 'moral progress occurs when benevolence is stretched to cover people whom those exercising power have never really thought of as members of the moral community' (Rorty 2006). This conception of moral progress dominates our thinking over the years. Darwin, for instance, held such a view, claiming that 'man's moral development' consists in a continual expansion of his 'social instincts and sympathies' (Stone 1974). Assume for the sake of argument that benevolence figures among the set of green virtues.[5] We would then need to show that the truly benevolent agent acts benevolently towards all morally considerable beings, members of future generations included. That is, we would need an account of the virtues, according to which, it is reasonable to expect agents to recognize diverse opportunities for the expression of a character trait like benevolence. The same would then apply to the other green virtues. This is the most general way to characterize the view of moral progress defended here. But, despite its currency and intuitive appeal, there is a contrary view – 'situationism' – that is now gaining many philosophical adherents. On this view, we progress morally precisely through a kind of *fragmentation* of our moral vision.

Some social psychologists and their philosophical followers[6] have argued that the attribution to agents of robust character traits is an error. Rather than being the product of enduring dispositions, these researchers suggest that our behaviour is significantly determined by the variables of the situations in which we find ourselves (Hartshorne and May 1963; Milgram 1969). The situationist attack on the virtues is complex but one of its key claims (for our purposes) is that agents do not display and should not aspire to display substantial trait consistency across diverse objects of moral concern. So the claim is both descriptive and normative. I have no quarrel with the descriptive claim, but I think the normative claim is unsound. In this section, I will therefore argue in favour of an *expansionist ideal*: that moral progress consists largely in an agent's recognition that character traits have substantial cross-situational applicability.

Situationists are not necessarily eliminativists about the virtues (though some are). John Doris, for example, thinks we can still talk about 'narrow' or 'local' traits, those which are confined very tightly to context. He thinks such traits can, 'underwrite very substantial behavioural predictability in their narrowly specified domains' (Doris 2002). If one is especially temperate in the gustatory domain, then one would do best to continue to be temperate there consistently, refining one's ability to avoid contrary-to-virtue temptations as they arise. One should not, however, expect to find new domains for this trait's application, the sexual domain for instance. Indeed, as far as progress is concerned, the idea seems to be that most of our moral failures occur when we attempt to expand our moral vision beyond these contexts. So on this view, the expansionist ideal is not just a false theory of moral progress, it is the most potent threat to moral progress. To avoid moral failures brought on by over-extension of the virtues, we should seek only to avoid 'near occasions for sin' (Dorris 2002, 147).[7]

This deeply deflationist view of moral progress should be resisted, for three reasons. The first is that it confounds our intuitions. Let us consider two virtues – respect and courage – in support of this claim. In her critique of situationism, Julia Annas asks us to think of how we would respond to someone, Mary, whose behaviour towards her colleagues was respectful but who treats waiters, shop assistants and football coaches with contempt. If I am one of her colleagues, how should I react to this disparity once I notice it? The situationist would have us see Mary as merely 'complex and intriguing', but Annas argues that nothing we have discovered forces us to change our negative attitude towards her (assuming this is how we react):

> [O]nce we see the whole picture, we don't just think of Mary as a person who happens to have different local traits; ... we re-evaluate her reasons in the original situation ... I cannot now think of her behaviour at work in the same way I did before. I now know that she is respectful to people whose opinions she needs to take account of, and disrespectful to people whose opinions she can safely ignore. (Annas 2005, 640)

In particular, the discovery of her rudeness might force me to think that Mary's behaviour towards me and other colleagues expresses hypocrisy or deference rather than respect. As Annas puts it, respect 'is not a trait that switches off in situations where the opinion of the people concerned can be ignored'. On the contrary, respect towards others is constituted in part by the recognition that such factors are strictly irrelevant. Annas' conclusion is exactly right: our considered judgement of Mary is not that she is respectful in context, but that she is not at all respectful. If this is right, there is something deeply suspect about the idea that respect is a highly context-dependent trait. This is not of course to say that there are no constraints on possible objects of respect. But Mary seems to have made a mistake in her expression of the trait and the mistake involves a failure to expand her moral vision adequately.

What about courage? For the situationist courage, like any other trait, will have many faces, all of them highly attuned to context. For example, we should not expect the person of physical courage to be morally courageous, nor should we expect moral courage to be consistently displayed towards 'diverse' objects of moral concern. Courage in the face of a police interrogator's threats does not, and should not reasonably be expected to, predict for courage in the face of a spouse's overweening demands (Doris 2002, 64–5). Nor, presumably, is there any normative connection between acting courageously to protect the interests of those near and dear and doing so on behalf of members of future generations.

On the other hand, Montaigne captures the expansionist position nicely:

> A man who is truly brave will always be brave on all occasions ... If he cannot bear slander but is resolute in poverty; if he cannot bear a barber-surgeon's lancet but is unyielding against the swords of his adversaries, then it is not the man who deserves praise but the deed. (de Montaigne 1991, 378)

Doris argues that an ideal like Montaigne's is pathological, and that we should instead opt for 'substantial behavioural inconsistency' which although it 'may confound our interpretive and ethical categories ... may also signal sound mental health' (Dorris 2002, 65).

Which normative ideal is better at signalling sound mental health? Take the case of Tim O'Brien, a conscripted infantryman who served in Vietnam 1969–1970. He relates the story of a girl in his fourth-grade class who was dying of cancer. She wore a headscarf to hide her balding head and was ridiculed for this by her classmates. Says O'Brien:

> I should've stepped in: fourth grade is no excuse. Besides it doesn't get easier with time, and twelve years later, when Vietnam presented much harder choices, some practice at being brave might've helped a little. (Miller 2000, 64)

O'Brien's report involves a specific view about moral progress, namely that we progress *through* ever-increasing expansion. Thus, the report shows us an agent who has made substantial moral progress precisely by recognizing his own previous failures of connection. We therefore presume that, other things being equal, he will in future better recognize opportunities for courage for what they are. Further, although O'Brien is talking here about moral courage, he might just as plausibly have been talking about some other virtue – benevolence, justice or compassion for instance. So the expansionist ideal is generalizable across the virtues. The situationist cannot recognize O'Brien's report, to the extent that it expresses this ideal, to be anything more than a kind of pathology. That, however, is patently false to our sensibilities in cases like this. If this is right, situationism cannot capture or explain a key ingredient of our concept of moral progress. When Doris says that the situationist ideal 'confounds our moral and interpretive categories', this is the sort of upheaval we should expect. At the very least, we should be wary of such counterintuitive re-interpretations of our practical lives.

The second reason to resist the situationist ideal has to do with the theory's inability to explain past episodes of moral progress. If in the past it had been fully embraced as a normative ideal, how could we humans have made the moral progress we evidently did make? If we were living in a slave-holding society, what could have prompted us to see slaves as entities with moral standing? In a patriarchal society, how would the situationist ideal allow us to see that women are not beyond the moral pale? In a society that routinely tortures and degrades non-human animals, how does this ideal permit space for the notion that such entities are morally considerable?

Again, to the extent that the situationist ideal forces us to look carefully at the minute social determinants of behaviour, it is valuable. It can allow us to see that we are often guided by situational pressures – expressed in group norms – that we may not, on reflection, endorse. The problem is that it does not give us scope for precisely this sort of reflection in really difficult cases. In such cases, when we step back to assess the situational pressures acting on us, we need to measure the demands they are making against a standard not itself derived from the situation. Instead, the standard is supplied by reason or a stable character disposition, or whatever. To revert to an earlier theme, standards provide the kind of *inflexibility* of character we need to avoid being swept up in conventional thinking and practices. The flexible agent, by contrast, is morally rudderless in the face of these forces. All the major advances in our moral vision required the ability to step back and assess current practices, showing that they are contrary to rational or virtuous ideals. There is something *revolutionary* about this process that the situationist ideal of merely avoiding near occasions for sin cannot capture.

In other words, the situationist ideal is simply not ambitious enough when we are faced with profound moral crises. This brings us to the third reason for rejecting that ideal: the expansionist ideal is *required* in order to meet the specific challenge posed by climate change. Here is an example that might get this thought going. What are we to make of the Nazi camp guard who spends his day assaulting, degrading and otherwise brutalizing prisoners, but who come evening treats his family members with apparent care and concern? Surely that he is not *at all* a caring or solicitous person, despite appearances to the contrary. Now, if our actions effectively treat the members of future generations as though they lack moral standing, then how exactly are we different from the Nazi? Of course, we do not go to work and torture people. But we make choices – mainly about what and how much to consume, but also about which people should run our countries – whose causal implications are now clear. Most of these consumption choices are linked to the burning of fossil fuels, and they are all therefore implicated in the dangerous increase of atmospheric concentrations of greenhouse gases, with *predictably* dire consequences for members of future generations. So we are evidently failing to expand our moral vision just as surely as the Nazi did and that is our chief moral failing at the moment.

Who exactly is this 'we'? It is tempting to say that it is all of us, the entire present generation of humanity, with those of us in the developing world justly receiving the lion's share of opprobrium. There is some truth to this, but my purpose here is not to apportion blame for climate change.[8] Instead, I want to put some flesh on the analysis offered so far by investigating how it might apply to a particular subset of humanity in the developed world, a group whose actions on the climate change front have come in for severe criticism from the rest of the civilized world since Kyoto. If some, or all, of the following claims generalize to citizens of other countries, so much the better.

## 3. Canada's climate failure

In evaluating the Nazi camp guard, we go beyond pointing out the badness of his treatment of the camp's prisoners and the objective harms he causes. We also want to shake him into grasping the fact that there are no morally relevant *differences* between the prisoners and his family members. Now consider the bulk of Canadians, enjoying comfortable middle class lives. In spite of what we might like to think of ourselves, the presumption that we are just, temperate, benevolent and mindful people is no longer well supported. Just as we were inclined to deny that Annas' Mary is genuinely respectful in *any* context – preferring instead to re-evaluate her actions towards us as displays of mere deference, for example – we Canadians must now be prepared to re-evaluate our own characters. This is why self-scrutiny is the key first step to moral progress.[9] But this is a tough work for us because we Canadians have generally seen ourselves in a broadly humanitarian, internationalist and socially progressive light. Indeed, there are two quintessentially Canadian activities: beating the Russians in hockey and occupying the moral high ground *vis-à-vis* our benighted American cousins. So how did we come to be international pariahs on the issue of climate change, an emissions rogue?

Although it ratified the Kyoto Accord, Canada was never serious about implementing the sorts of policies that would have allowed it to meet its emissions reduction targets. At Kyoto, Canada committed to reducing its greenhouse gas emissions 6% below 1990 levels. As of 2005, its actual emissions were 24% above 1990 levels or 30% above the target. This is the world's worst Kyoto shortfall. Much of this failure can be attributed to sheer indifference on the part of liberal governments led successively by Jean Chrétien and Paul Martin. But with the rise to power in 2006 of the Conservative Party, environmental indifference turned to sheer hostility. Just prior to becoming prime minister, Conservative Leader Stephen Harper described Kyoto as a 'socialist scheme' for transferring wealth from rich countries to poor ones. The Conservative power base is in the province of Alberta, home to the infamous oil sands. And no sober person is the least bit skeptical about Harper's intention to grow that industry dramatically, no matter what environmentalists, or anyone else, says in protest.

Alberta contains vast quantities of bitumen, a sulphur-rich substance that requires upgrading to synthetic crude oil. There are between 165 and 173 billion barrels of proven reserves, those which are extractable given current technology. This is second only to Saudi Arabia in proven reserves. Mining the oil sands is fantastically energy intensive because in its natural state it is too viscous to pump. So it needs to be removed from the ground and then have its sand and oil components separated from one another, thus allowing it to flow. Between 20 and 34 $m^3$ of natural gas are required to produce a single barrel of bitumen. Currently, the oils sands uses about 21 million cubic metres per day of natural gas, an amount projected to rise to 60 million cubic metres per day by 2015. Greenhouse gas emissions created through extraction and upgrading of bitumen are a stunning 3.2–4.5 times what they are for conventional oil. Syncrude, the largest company working in the oil sands, emitted 10.3 million tonnes of greenhouse gases in 2005, an amount equivalent to emissions from 2.7 million cars over the same period (Draper and Reed 2009, 450–2). As the Pembina Institute states in its 2008 oil sands 'report card', the technologies involved in oil sands production 'make the product among the most environmentally costly sources of transport fuel in the world' (Draper and Reed 2009, 451–2).

But, of course, no government can get away with exploiting such an unpopular resource without some attempt to spin its way clear of criticism. So the Conservatives engage in half measures. They run public awareness campaigns encouraging citizens to be more energy efficient, impose voluntary or intensity-based targets on industry and rely on selected technology subsidies. These measures are neither cost-effective nor environmentally effective (Snoddon and Wigle 2009). As Simpson, Jaccard, and Rivers put it 'the only policies that will work place a price on greenhouse gases or regulate their emission' (2007, 250). So why has the government refused to take this sort of action? Two answers are usually given.

The first suggests that a carbon tax would cause economic ruin, not just for Alberta but for the country as a whole. But this claim has been contradicted in a study recently commissioned jointly by the David Suzuki Foundation and the Pembina Institute. The study calculates the cost to the Canadian economy of striving to meet an emissions reduction target of 25–40% below 1990 levels by 2020. The report shows that if it puts a serious price on carbon, substantially increased production of renewable energy and captured and stored the $CO_2$ from the Alberta oil sands (as well as from power plants), Canada could meet this target. And it could do so while the economy enjoyed a 23% *growth* in GDP from 2010 to 2020, just 3.2% less than GDP growth under a business as usual scenario. It would also see the creation of close to 2 million net new jobs in this period (Bramley, Sadik, and Marshall 2009, 4–6). To call this state of affairs, economic chaos is transparently mendacious, but the report was nevertheless summarily dismissed by the governments of both Alberta and Canada for just this reason.

The second line of defence sidesteps the question, arguing in effect that there is no need to put a price on carbon. Even with the oil sands running at full tilt, so goes the argument, Canada is such a small country, responsible for just 2% of global emissions. So we really are not doing that much damage to the atmosphere. The response disingenuously suggests there are two *necessary* conditions for continuing to develop the oil sands: doing so makes us rich and is relatively harmless. This entails that if either of these two conditions was not met, the oil would stay in the ground. But while this is obviously true of the first condition, it is extremely naïve to believe it is true of the second. Does anyone really suppose that the current government of Canada would abandon development of the oil sands if the process were responsible for 50% or 80% of global emissions?

In any case, though the sands are egregiously polluting, pointing to the emissions created in extracting and upgrading bitumen is a potentially dangerous distraction from the real issue. Many energy analysts are now arguing that we have reached, or are very close to reaching, peak conventional oil (see, e.g. Rubin 2009). If this is right, and given the ever-increasing demand for oil, there will be a supply shortfall in the near future. And Canada is poised to meet this shortfall. In 2008, the oil sands were producing one million barrels per day, a number that is expected to rise to four or five million barrels per day by 2020 (Jaccard 2009, quoted in Homer-Dixon and Garrison 2009, 114). All of this is possible given the quantity of proven reserves in the oil sands. But the reserves are just a small fraction of the larger resource, which has been estimated at *1.7 trillion barrels*. To put this amount in perspective, it exceeds the combined proven reserves of all oil producing countries on the planet by approximately 500 billion barrels. The cost of extracting this resource means that the bulk of it will likely stay in the ground, but if even a very small portion of it can be extracted, Canada could become *the* world's oil superpower. I believe our current leaders are drunk on this prospect, and it explains why they are seen by many to be engaging in attempts to undermine or dilute international agreements on emissions reductions.

So development of the sands proceeds apace. In January 2010, significant new projects were announced by ConocoPhillips, Total SA, Husky and BP, effectively ending the temporary recession-driven lull in oil sands development. In the words of Jean-Michel Gires, president of Total Canada, 'our bet on the oil sands is definitely for the long term' (Conoco 2010). And just in case the USA, currently the world's largest importer of Canadian oil, decides it does not want this oil – a distinct possibility since there is a movement afoot in some states to ban it or place heavy tariffs on it – the Canadian government and private industry are investing very heavily in pipeline construction projects to bring the oil to other countries. A single one of these pipelines will bring up to 525,000 barrels of oil a day from northern Alberta to the Pacific coast, for shipment to the markets of Asia and beyond. Just days before the Copenhagen conference in December 2009, Stephen Harper conducted a brazen, whirlwind tour of this part of the world – including stops in both India and China – telling anyone who would listen that we have got the resources to power your burgeoning economies.

The 2% argument indicates clearly that Canada is behaving like a reckless herdsman on the commons, seeking to perpetuate a system of economic relations that is destructive and unsustainable. As perverse as it sounds, that is why Canada *wants* the world to focus atomistically on its (relatively paltry) domestic emissions. Canada is an oil producer with a small population. Even if it became the sole supplier of the world's oil, it could continue to make a similar argument because it consumes a relatively small amount of the stuff itself (so long as we do not calculate this on a *per capita* basis). But this counterfactual scenario reveals the absurdity of the claim: Canada is simply not counting the emissions released when the oil it exports is burned downstream. Indeed, if one were to calculate emissions the way Canada prefers, one could dismiss the emissions contributions of the world's five largest oil exporters. After all, these countries – Saudi Arabia, Canada, Russia, Iran and the United Arab Emirates – are collectively responsible for just 10% of the world's emissions (and just half that if you exclude Russia). The problem, of course, is that their combined oil exports total nearly 24 million barrels a day. If Canada and its dubious allies are successful in convincing a critical mass of other nations to play the game the way they want, the result will therefore be the total ruination of the atmospheric commons. The slide from the tipping point to the point of no return will be assured.

As we have seen, perhaps the best way to get out of the trap of calculative thinking that underpins this stance is to think in terms of the virtues. In the light of the direction our government is taking us, it is thus high time for Canadians to decide what sort of people we believe we are. The key to any solution to climate change lies in 'engaging intergenerational motivations', as Gardiner puts it. And one way to do this is to imagine 'the potential effects of abrupt change on future people' (Gardiner 2009, 156). Think of a future person whose world has been blighted by climate change and put yourself for a moment into his or her shoes. Suppose that person could somehow address members of our generation. 'Do you really mean to tell me', this bewildered person might say to a Canadian of the early twenty-first century, 'that you refused to take emissions reduction measures that would have reduced your GDP by 3.2% over a 10-year period relative to business as usual because such measures were deemed too *costly*? Even though your economy would still have grown 23% in this period? What mind-boggling greed!'

If this image makes us uncomfortable, it is because it casts us as vicious people. Are we? We certainly do not think so. For example, in December 2009, evidence emerged that Canadian forces currently stationed in Afghanistan had been transferring prisoners to Afghani authorities, whereupon many of these prisoners were tortured. It was revealed that the Canadians knew this would likely happen but did nothing to protect the prisoners. The government then tried to conceal the relevant evidence. Canadians were outraged that their government appeared to be colluding in torture. Fast forward again to our future person. She might ask us why we reacted this way. Our answer would likely be that those prisoners have moral standing on a par with every other person.

We might add that no just or benevolent or mindful person would countenance this sort of treatment. 'Really?' our future critic might respond, 'And what about us? In spite of, or perhaps because of, some of your economists' attempts to "discount" our interests, here we are – with monumental existential challenges.[10] Do we lack some morally essential property? What could it be? Perhaps it is the fact that at the time you made the decision to abandon efforts at climate change mitigation we did not *yet* exist. But is this a morally relevant property, given the fact that your actions clearly had the power to bring into being a world that was better than this one? In condemning your government's actions on the Afghani detainee issue, you refused to take the geographical remoteness of the prisoners – not to mention their dark skin or foreign habits – as reasons to deny their moral standing. You effectively claimed joint membership with them in a moral community. And yet in virtue of the causal connection between your actions and the state of our world, *we too were members of your moral community* when you made your fateful choices. We are genuinely baffled, to put it mildly, that you could you not see this'.

Thought experiments like this allow us to see that our failure lies, precisely, in neglecting to *expand* our moral vision in a manner clearly dictated by the evidence. And the conclusion of this exercise must be that, because of this failure, our reaction to the mistreated Afghani prisoners does not in fact reveal us to be just and benevolent *people* at all. Our dispositions and moral vision are too fragmented to warrant that self-description. This, I think, is what Canadians need to say about themselves in view of the governments they continue to elect. If they do not, then their fulminations about the rights of Afghani prisoners can be dismissed as self-righteous hot air. Canada is among those countries currently impeding the moral progress of the species. Indeed, its actions may help bring us closer to a Hobbesian hellhole in which the species not only fails to progress but also regresses dramatically. We are actively undermining the vital interests of a totally defenceless group of humans, and that is one meaning of the idea that the people of Canada are becoming, in Monbiot's phrase, brutalized by oil.

## Notes

1. *Climate change: Global risks, challenges and decisions*, University of Copenhagen, March 2009 (http://climatecongress.ku.dk/speakers).
2. Here I employ the notion of 'comparative advantage' in a non-technical sense, though what I have to say about it is related to Ricardo's technical use of the term.
3. Other philosophers have called for a virtue-ethical approach to environmental ethics more generally. See, for example, Sandler and Cafaro (2005), Sandler (2007) and Van Wensveen (1999). The classic in this literature is Hill (1983).
4. This is not to claim that there are no members of the present generation who will be harmed by climate change. But the major climatic disruptions are likely to occur in a generation or two, and so I focus on this group of people.
5. My task in this paper is not to put forward a list of green virtues appropriate for climate change. The ones I refer to – justice, benevolence, sympathy, respect – are fairly obvious candidates for this role, but a fuller defence of the claim that they are in the appropriate set would require a separate treatment.

6. The most sophisticated philosophical proponent of situationism currently working is John Doris, on whose account my analysis focuses. See Doris (2002); also, Doris (2005), Harman (1999), Harman (2008). See also Vranas (2005). Vranas' analysis bolsters the conclusions of professed situationists like Harman and Doris.
7. In a similar vein, Harman argues that we should not strive to change people's characters. Instead, we should craft our social institutions so that the sorts of situational pressures that give rise to moral failures rarely happen. See Harman's website paper cited in Note 6.
8. Others have done so ably. See, for example, Singer (2002); Garvey (2008), chap. 3; Schönfeld (2011).
9. As Moody-Adams (1999) argues.
10. It is important to characterize the challenges as existential so as to forestall objections arising from Parfit's non-identity problem for future generations. The problem is that specific members of future generations cannot be said to be harmed by our actions since *they* would not have existed but for those actions. So long as non-existence would not be better for them they have nothing to complain about. Parfit himself addresses the problem by denying that any future person is harmed by us, but that the world we bring about is objectively worse than the one we might have brought about, for which we can be blamed. See Parfit (1986, part 4). I prefer to say that in the world we are in the process of bringing about suicide will be a live option for huge numbers of people. That is, I think that many people *will* likely prefer non-existence to existence in the world we are creating for them. And even if they cannot (for whatever reason) commit suicide, they might still consider non-existence preferable to existence in this world. A chilling portrait of a post-apocalyptic world like this, a world marked by mass suicides, is in Cormac McCarthy's (2006) *The road.*

## References


Bramley, M., P. Sadik, and D. Marshall. 2009. *Climate leadership, economic prosperity: Final report on an economic study of greenhouse gas targets and policies for Canada.* Vancouver, BC, Canada: David Suzuki Foundation and Pembina Institute.

Doris, J. 2002. *Lack of character.* Cambridge: Cambridge University Press.

Doris, J. 2005. Replies: Evidence and sensibility. *Philosophy and Phenomenological Research* 71, no. 3: 656–77.

Draper, D., and M.G. Reed. 2009. *Our environment: A Canadian perspective.* Toronto: Nelson Education.

Gardiner, S.M. 2009. Saved by disaster? Abrupt climate change, political inertia and the possibility of an intergenerational arms race. *Journal of Social Philosophy* 40, no. 2: 140–62.

Garvey, J. 2008. *The ethics of climate change: Right and wrong in a warming world.* New York: Continuum.

Hansen, J., M. Sato, P. Kharecha, D. Beerling, V. Masson-Delmotte, M. Pagani, M. Raymo, D.L. Royer, and J.C. Zachos. 2008. *Target atmospheric $CO_2$: Where should humanity aim?* http://arxiv.org/abs/0804.1126.

Harman, G. 1999. Moral philosophy meets social psychology: Virtue ethics and the fundamental attribution error. *Proceedings of the Aristotelian Society* 99: 315–31.

Harman, G. 2008. *Skepticism about character traits.* www.princeton.edu/~harman/Papers/index.html.

Hartshorne, H., and M. May. 1928. *Studies in the nature of character, I: Studies in deceit.* New York: MacMillan.

Hill, T.E., Jr. 1983. Ideals of human excellence and preserving natural environments. *Environmental Ethics* 5: 211–24.

Jamieson, D. 2007. When utilitarians should be virtue theorists. *Utilitas* 19, no. 2: 167.

Kartha, S., T. Athanasiou, P. Baer, and D. Cornland. 2009. *Cutting the knot: Climate protection, political realism and equity as requirements of a post-Kyoto Regime*, 4. http://www.ecoequity.org.

Lovelock, J. 2009. *The vanishing face of Gaia: A final warning.* London: Allen Lane.

McCarthy, C. 2006. *The road.* New York: Knopf.

McKinnon, C. 2009. Runaway climate change: A justice-based case for precautions. *Journal of Social Philosophy* 40, no. 2: 187–203.

Milgram, S. 1969. *Obedience to authority.* New York: Harper and Row.

Monbiot, G. 2009. Please, Canada, clean up your act. *The Globe and Mail*, November 30, A15.

de Montaigne, M. 1991. *The complete essays.* Harmondsworth, UK: Penguin.

Moody-Adams, M.M. 1999. The idea of moral progress. *Metaphilosophy* 30, no. 3: 168–85.

Parfit, D. 1986. *Reasons and persons.* Oxford: Oxford University Press.

Rorty, R. 2006. Philosophy's role *vis-à-vis* business ethics. *Business Ethics Quarterly* 16, no. 3: 369–80.

Rubin, J. 2009. *Why your world is about to get a whole lot smaller: Oil and the end of globalization.* Toronto: Random House Canada.

Sandler, R. 2007. *Character and environment.* New York: Columbia University Press.

Sandler, R., and P. Cafaro, eds. 2005. *Environmental virtue ethics.* Lanham: Roman and Littlefield.

Schönfeld, M. 2011. Amerigenic climate change. In *Environmental ethics for Canadians*, ed. B. Williston, 283–90. New York: Oxford University Press.

Simpson, J., M. Jaccard, and N. Rivers. 2007. *Hot air: Meeting Canada's climate change challenge.* Toronto: Emblem Books.

Singer, P. 2002. One atmosphere. In *One World*, chap. 2. New Haven, CT: Yale University Press.

Snoddon, T., and R. Wigle. 2009. Clearing the air on federal and provincial climate change policy in Canada. *IRPP Choices* 15, no. 11: 1–26.

Stone, C.D. 1974. *Toward legal rights for natural objects.* Portola Valley, CA: Tioga Publishing Company.

Van Wensveen, L. 1999. *Dirty virtues.* Amherst: Humanities Press.

Vranas, P.B.M. 2005. The indeterminacy paradox: Character evaluations and human psychology. *Noûs* 39, no. 1: 1–42.

# Climate change and philosophy in Latin America

Ernesto O. Hernández

*Philosophy, University of South Florida, Tampa, FL, USA*

This paper aims at surveying the current philosophical issues concerning the climate change crisis in Latin America. The work attempts to analyze some central policies, particularly those that fostered economic progress in the region at the expense of human and environmental depletion. Historically, Latin America remained at the periphery of philosophical inquiry following the long standing multiple manifestations of colonialism. As a result, the systematic philosophical reflections about climate change in the region have been scarce at best. Here, I have addressed three main topics: political ecology, economic development, and indigenous issues. Latin American political ecology generally seeks to identify the ontological schemes necessary to mitigate, cope, and adapt to the climatic crisis. In addition, developmentalism played a significant role in exacerbating the state of Latin American political, economic, and ecological systems. Environmentalists all throughout the region strive to find alternatives to replace the failures of developmentalist policies. Consequently, indigenous concepts such as Sumak Kawsay or 'good living' offer alternate philosophical modes for Latin Americans, with their complex ethnic diversity, to derive their ontological and epistemological foundations on the problems posed by climate change. My work represents a sketch reporting the various inception points of different eco-philosophical views portraying legitimate instances of Latin Americans thinking critically about their own reality and surroundings. These may serve as alternate methods to forge models that could potentially contribute to universal philosophical concepts endowing humanity with broader epistemic tools to tackle the crisis at hand.

This essay is an introduction to the philosophical engagement with climate change in Latin America. It is not an exhaustive survey of the ideas, concepts, and views on climate and the environment in Latin thought. Its purpose is to take a preliminary look at the philosophical literature devoted to the topic of climate. I hope to show that issues formerly thought to be isolated are now converging on common ground, and suggest a framework for further research on climate-related philosophical pursuits in Latin America.

The environmental crisis known as 'climate change' has sparked the interest of activists and theoreticians. Dealing with climate change may well be humankind's most serious challenge. This challenge opens up new fields of inquiry within many disciplines. Climate change is the broadest topic of interdisciplinary relevance ever studied by the scientific community. It attracts economists, biologists, sociologists, physicists, and engineers alike. Philosophers, however, have largely neglected to engage with this new topic, and many of them still fail to address the urgent need to reflect about the existential questions of our present predicament. Climate change does not respect political boundaries or cultural differences. Arguably certain political powers and economical theories are the perpetrators of the crisis. Climate change is not only anthropogenic in nature, but also specifically the product of the industrialized West.

While highly developed countries are the major cumulative perpetrators, underdeveloped countries around the world are the most vulnerable to the consequences.

The philosophical establishment of Europe has often ignored (for reasons too vast to discuss here) non-Western and non-European philosophical activities, by deeming them 'non-philosophical'. As Jorge Garcia points out, North Atlantic philosophy recasts the model of Europe by inheriting the same self-proclaimed exclusivity to philosophize. My argument aims to bring up these two hitherto neglected topics, Latin American thought and climate change, to the attention of the philosophers in the North. That Latin Americans can contribute universal, epistemic, and ethical perspectives about a problem without geographical, biological, or political boundaries tells us something about the value of these non-standard approaches.

## 1. Latin America's environmental conditions and the problem of development

Latin America is already suffering from the effects of climate change. National populations are suffering from poverty, famine, scarce access to drinking water, and poor public health. The systematic exploitation of natural and human resources, required to feed the world's economies, is jeopardizing the ecological equilibrium of Latin America. In this context it is not an accident that Latin America is the continent with the richest biodiversity on the planet. This makes it extremely attractive for capitalism to continue the systemic depletion of the Latin American environment.

The Colombian philosopher and anthropologist Arturo Escobar proposes a framework for a Latin American political ecology. For him, 'Political ecology can be defined as the study of the manifold articulations of history and biology and the cultural mediations through which such articulations are necessarily established' (Escobar 1999, cf. 3). He pursues new ways to rethink Latin America's situation by questioning basic concepts such as development, modernity, and progress. Turning to the distinction between humans and nature as conceived by neoliberalism, he states,

> Historically, the production of surplus with the concomitant social and institutional differentiation allowed humans to emancipate themselves from nature, albeit at the price of enslaving part of the population. (Escobar 1999, 6)

While some portions of the global population reach a high standard of living, they do so at the expense of others, making others suffer the downside of an invidious economic transaction. The costs of such a high standard of living not only affect people but also impose an unsustainable burden on ecosystems. For Escobar the task of political ecology is 'to outline and characterize these processes of articulation and its goal to suggest potential articulations realizable today and conducive to more just and sustainable social and ecological relations' (Escobar 1999, 4). The first step of doing so is the reconceptualization of the relation of humans with nature. Such efforts must come from the effort to reframe an 'anti-essentialist' view of nature, a view that rejects the dualistic ontology of conventional Western thought that conceives of self or mind as being detached from body or nature.

Escobar pursues a middle ground between two ontological views about nature. One view is that nature possesses immutable properties independent of human perception. Another is that the concept of nature is produced by history, culture, and social constructions. Escobar attempts to marry the two views by acknowledging, on the one hand, 'the existence of a biophysical reality—prediscursive and presocial' (Escobar 1999, 1). On the other, he acknowledges that to an extent our conception of nature is 'artificially produced' (Escobar 1999, 2). He argues that this entails 'an unprecedented ontological and epistemological transformation which we have hardly begun to understand' (Escobar 1999, 2). The difficulty for him lies in the structures of these models, their incompatibility with other structures, i.e. non-dualist, non-essentialist ontologies, and epistemologies.

Such reorganizations entail the emergence of a different approach by which these epistemologies interact with each other. Escobar points out that these epistemologies and upstream ontologies that support them are not in any way negotiating stances that may lead to a synthetic practical/political outcome. Dominant epistemologies in the West are 'vertical' structures that are hierarchically arranged. In them, the subject trumps the object, culture rules nature, and individualism champions the collective.

For Escobar it is imperative to de-essentialize the dominant epistemologies in order to make them interact more effectively with local, subaltern or 'peripheral' epistemologies and upstream ontologies. These 'anti-essentialists' and 'post-constructivists' epistemologies and ontologies break away from rigid conceptions about nature. This suggests a revaluation of the human–environment relations.[1] The proposed de-essentialization aims to promote what Escobar calls a 'flat relational framework': multi-directional channels that facilitate interfaces between previously incommensurate notions of reality.

For instance, after WWII, and particularly under the Kennedy administration, an aggressive plan to 'modernize' Latin America, Africa, and Asia came about. This era laid the groundwork for the project of development for the past half a century. The project of development, as the social philosopher and cultural analyst Saldaña-Portillo points out, is 'inextricably linked to managing a crisis in capitalist production precipitated equally by the exhaustion of colonial capitalism's expansive capacities and by the greatly expanded productive capacity of the U.S. postwar economy' (Portillo 2003, 19). The reconstruction of Europe after WWII left the USA in need to create and secure markets to warrant their economic prosperity. Latin America was the obvious choice to secure cheap resources and human labor. With the promise of development, progress, and modernity, the USA launched programs such as *Alianza para (el) progreso*, which allocated aid to help in the industrialization of Latin countries. During this period the USA singlehandedly established stronghold institutions that secured their investments in the new post-colonial world order.

While these projects were advertised as efforts to mitigate poverty and misery on a continent that had suffered systematic exploitation over centuries, they also contributed to massive problems. Armies of professionals from the developed world were deployed to assess the conditions of these non-developed countries. Their evaluations led to the creation of more regional projects of modernization. The process of modernization became synonymous with industrialization. In effect, development is the process by which poor regions become industrialized in order to progress toward modernity. Progress was measured on the basis of resource and human labor exploitations. The choice for progress, which was offered to Latin America, was a choice at the expense of its natural assets and human resources.

The notions of modernity, development, and progress served to impose *the way of life* of the 'experts.' Modernization projects were developed through the force and authority of expert knowledge, sometimes against the best interests of the regions in welfare and livelihood. The whole process also deepened the perception of developed countries to act with a sense of superiority and, at the same time and almost paradoxically, to insinuate that the underdeveloped countries are not capable to progress on their own. As Saldaña-Portillo rightly points out, 'Even as development emerged in concert with the universal right to self-determination, it nonetheless carried within it the traces of imperial reason, of an evolutionary hierarchy and racialized subordination' (Portillo 2003, 22). The US-led project was aimed to shield Latin America from the communist threat. After the Cuban revolution in 1959 the USA grew even more concerned about the spread of Marxism across Latin America. Thus, the projects of development came about with a new form of dominance. The approach was set to impose the set of values and the work ethic of capitalism, while at the same time blocking any framework, local or from foreign sources. Another consequence was a new emphasis on individualism and subjectivity. Saldaña-Portillo notes that 'the development imperative once again aims for the interiority of subjectivity: some men must be made ready to be ready to become the risk taking, innovative subjects of capital, while others must be made ready to be ready to become the disciplined subjects of monotonous wage labor' (Portillo 2003, 31). Latin America faces the need to address the intensification of neoliberalism, together with a renewed promise of progress through the clean energy projects promoted by developed countries. There is a danger of falling prey to the same illusions that the dawn of developmentalism offered. Chief among them is the idea that 'the proper application of development aid in the fields of "scientific advances" and "industrial progress" would rapidly remake the world in the image of the United States' (Portillo 2003, 23). Actually, however, development brought about more prosperity for those already developed at the expense of the vulnerability of the underdeveloped. In the 2011 State of the Union address President Obama stated, 'This March, I will travel to Brazil, Chile, and El Salvador to forge new alliances for progress in the Americas'. Both in Chile and Brazil negotiations about clean energy deals are expected to take place. In the case of Chile the conversations focus on nuclear energy, while in Brazil conversations will target biofuels and fossil fuels. Reminiscent of the projects of development in the 1960s, the USA is using this renewed effort of development to create political and economical leverage against China.

The paradigm of development, in all its complexity, awakens the need for these 'developing' countries to realize an alternative that would satisfy basic social needs and sustainably improve living conditions. The vision of a sustainable alternative suggests that Latin America needs to move beyond the paradigm of development. This move would aim at the creation of a new paradigm, from within Latin American circumstances, needs, and interests. Escobar suggests a post-development political framework in which local epistemologies and ontologies may function according to the needs of the inhabitants and their environment. Such structures are envisioned to stand on their own but, at the same time, to be sensitive to the adaption and integration of useful features of other frameworks.

The debates on climate change in developed countries are focussed on the dual goal of reducing carbon footprints and mitigating the ill effects of greenhouse gas (GHG) emissions. It is commonly agreed that an increase in global mean temperature of more than two 2°C would have catastrophic consequences to the ability of the Earth system to sustain terrestrial life as we know it. Underdeveloped countries made minimal contributions to these dangerous emissions, yet they are seeing the worst effects. Underdeveloped countries are the victims, developed countries are not.

Developed countries seek solutions by *future* net reductions of their GHG emissions while maintaining the same quality of life and standard of living. This schema, their representatives argue, should become a universal model. But this is not persuasive. Little has been accomplished in the political arena to overhaul and curtail the practices that result in GHG emissions. Collectively developed countries have the outlook that imposes an economical value to nature. The rubric of political discourse does not account for the intrinsic value of nature. It is only based on its value relative to human interests.

The ethos that emerges from industrialization, a consumerist and ever-progressive attitude, amounts to the exercise of control over nature but also over other humans. New forms of sustainable technologies of industrialized nations can put strain on other cultural identities, reviving once again the specter of colonialism. For instance, conventional goals of sustainable development in wealthy countries are to keep the same standard of living and to maintain the same economic growth. If this requires the conversion of vast swaths of rainforest (inhabited by native tribes) to sugar cane farms for biodiesel production, so be it.

Developed countries to date have shown little willingness to cut back on their rampant consumption. Recent proposals by environmental thinkers in the developed world, such as by the British journalist G. Monbiot, to drastically cut back emissions seems unrealistic due to the inability to imagine a simpler lifestyle that amounts to a sacrifice of individual freedoms. Instead the action of choice seems to be the push to create a massive increase in energy production to prolong the unsustainable lifestyles of consumption and wastefulness for a few more years.

A further issue in developed countries is the fashion among the well-to-do to shift consumption to 'green products' that offer no substantial, long-term solutions to the climate crisis. This consumption shift rather seems to serve as a cathartic effect, relieving individuals from feelings of guilt and giving them the sense that they are 'at least doing something'. It is a superficial, anthropocentric view that veils the problem with ineffective, insufficient, or illusory solutions. These collective and individualistic notions, which are not adequate to address climate change, emerge from the particular circumstances of developed countries.

Perhaps these collective and individualistic responses of developed countries are suited to meet *some* of their needs. But they do not reflect the needs and circumstances of underdeveloped countries. The 'green' models proposed by developed countries are disconnected from the environmental and social conditions of Latin American. Nonetheless, and even though they do not address Latin American circumstances, these 'green outlooks' have found their way to influence some its environmental movements.

The Uruguayan sociologist Eduardo Gudynas casts Latin American environmentalism as an array of conceptual and practical positions. He identifies them as 'multiple greens'. They encompass activists as well as academics, with differences so sharp that their only common ground is environmental protection. Typical of some of these movements is the association of environmentalism with a concern for social problems. This concern is characteristic of Latin American greens and different from first world environmentalism. Highly developed nations have contributed to the depletion of resources in Latin America without suffering the domestic human consequences of such devastation. This is why citizens in developed countries tend to think about climate change as a future emergency instead of an ongoing process of social and natural degradation. From the perspective of Latin American environmentalism, the climate crisis is not a problem looming in the future, it is already a reality.

Gudynas singles out the two most questionable Latin American green positions. The first is what he calls the 'neoliberal environmentalism' (Gudynas 1992, 110). This is a movement that 'promotes a posture of conservation of the natural resources, not by their intrinsic values, but because they are useful to economic progress' (Gudynas 1992, 110). This approach treats nature as a 'bank' or a 'basket of goods' whereupon the conferred value is measured by its utility. According to this approach natural resources must be sustainably managed to ensure the continuous growth of the Western World's capitalist economy. Nature is conceived as a means to an end, a phenomenon that ought to be controlled, manipulated, and exploited for the benefits of the few. This approach is conceptually disconnected from the socioeconomic situation of the continent. Neoliberal environmentalists respond to the climate crisis with a push for new technologies to reduce the carbon footprint of the market.

Governments of countries such as the USA, the global hotbed of neoliberalism, do not question the economic model of eternal growth and accordingly do not explore the possibility of mitigating climate change through the implementation of alternative economic models. They resist an environmental paradigm change and remain advocates of the same system that created the crisis in the first place. Their solution is to pursue minimalist changes such as the adaptation of green technologies. This is not to say that research and development of these technologies is not valuable, but this is to suggest the possibility that the system itself has failed. But the neoliberal environmentalism of Latin America that takes its cue from the USA projects the contrary vision; in Gudynas' words (2009, 40), it 'generates the illusion of a "benign capitalism". The fundamentals of its functioning go unquestioned, as do those of the commoditization of nature or the supporting roles of social programmes'.

Another problematic Latin American green position is what Gudynas calls 'ecomessianism' (Gudynas 1992, 111). He defines this movement as 'an unjustified exaggerated faith in environmentalist ideas as agents of change that will save humanity' (Boff 1995, 129). This movement is subjective, optimistic, and utopian. This type of attitude is a serious epistemological hurdle to establish a comprehensive critical stance to climate change. In actual practice, the boundaries between neoliberal environmentalism and ecomessianism blur; both share the conviction that climate change is an easily mastered problem. A blend of both positions expresses itself in technological messianism – the faith that technology will always save us.

Some conceive the technological approach as the *only* viable path to mitigating climate change. There are several reasons for this argument. The technological approach looks attractive because it does not challenge the conventional economic model, instead supports the conventional social behaviors such as consumerism. It also looks attractive because in supporting such behaviors, it erases the burden of responsibility for such lifestyles. The powerful nations of the North impose the technological approach on underdeveloped countries instead of joining them in a global effort of moving toward a sustainable economic system. This I believe is one of the worst symptoms of eco-imperialism (Driessen 2003, passim).

## 2. Disempowering Latin America

The imposition of technological solutions disenfranchises Latin American countries and makes it harder for them to contribute to efforts to find systemic solutions to the climate problem. The poor infrastructure and dated means of production on the continent handicap Latin America to pursue the necessary research and development of green technologies. This circumstance constitutes yet another layer of complexity in the dynamics of neo-colonialism. There are also practical problems that affect the ability of Latin American nations to deal with the climate crisis. Politically Latin America lacks an orchestrated effort to articulate its own set of approaches. Its politico-economical conditions are often interpreted as systemic failures by the international community (Ruatta 1996, 92).

If the climate crisis has only a technological solution, as neoliberalism suggests, then Latin America, along with other poor regions of the global village, will be silenced. Their efforts will remain at the periphery, and their policies will be reduced to passive dependence on developed countries to 'come up with the right plan.' The obvious problems with this technocratic approach are not only that other, indigenous, and arguably more effective solutions to the climate crisis are ignored, but also that it ails from a misperception of the entire crisis. For how is it possible to solve the problem with the same tools that created it?

How, then, would Latin America be able to act and to serve as a model of climate change mitigation? The continent's ethnic, cultural, political, and economical factors converge in Latin American climate ethics. The environmental crisis intensifies the struggles of oppressed groups seeking social justice. There are multiple ways by which different groups are constructing the discourse to pursue justice. For our purposes, we can distinguish softer approaches from more radical ones.

Softer ecological movements seek to re-establish the struggles through a perspective based on the normative ideas of sustainability and equal opportunity (Leff 1994, 43). This perspective concerns the social problems of environmentalism, such as extreme poverty, lack of resources, and unjust power relations. This approach, while an improvement over the *status quo*, ignores the deep problems that arise from anthropocentrism. By training its focus on human interests it neglects the remainder of nature.

A more radical approach proposes not only the establishment of an egalitarian system where humans enjoy the same standard of living and quality of life but also acknowledges an inalienable right of nature. At least three movements within Latin America have underpinned this radical approach: Eco-theology of liberation, Ecofeminism (Cova 2005, 42), and what I identify as Indigenous Pragmatism.

## 3. Eco-theology of liberation and ecofeminism

The movement known as liberation theology emerged in response to social injustice. This movement claims a hermeneutical interpretation of the Gospels by which God always sides with the poor and the oppressed against the wealthy and the oppressors. Although the movement originated in a small group of Latin American theologians it eventually spread all over the world, blurring political, social, and religious boundaries. One of the pioneers of liberation theology is the Brazilian Leonardo Boff. His theological and philosophical outlook is shaped by Meister Eckhart, Francis Assisi, Saint Augustine, among others. Boff earned his doctorate in theology under the supervision of Joseph Ratzinger, the later Pope Benedict XVI. His 'radical' ideas about the poor prompted the Congregation for the Doctrine of Faith, then led by his former teacher Ratzinger. Eventually Boff left the church after further conflicts with his clerical superiors.

Boff brought the notion of environmentalism to liberation theology. His first publication on this matter was *Ecology and liberation* (1993, translated to English 1995). The argument in favor of the poor and the oppressed is extended to nature and the environment. In Boff's words, 'Liberation theology is in communion with the political aspirations of many social groups that seek a society concentrated on the dignity of the human person and on a form of participation that, through labor, satisfies the basic needs of food, shelter, health, education, and leisure, and opens up areas of freedom for creativity and the collective building up of society' (Boff 1995, 126).

These conditions are also compatible, if not identical to the consensual requirements of behavioral change to mitigate the climate crisis. Liberation theology seeks an increase in quality of life in an ecologically sustainable way. Boff asserts: 'The best projects, practices, and organizations are those that do not aim exclusively at the quality of goods and services, but at the quality of life, in order to make that life truly human. Society as a whole should make a life of this kind its goal. The alliance that is in the course of establishment between men and women and nature, in terms of brother-sisterhood and veneration, also forms part of human well-being' (Boff 1995, 129). Boff stands as a pioneer in the inclusion of nature and the oppressed in the Latin American environmental discourse. He framed the debate that enabled marginalized groups, such as the poor, women, and indigenous populations, to participate and contribute to the Latin American discourse on social and environmental conditions.

Ecofeminists, on the other hand, highlight a deeper problematic by arguing that women, particularly in developing countries, find themselves trapped in gender inequalities. They claim that ecological movements have lacked the sensitivity to fully acknowledge and actively include the most affected groups. Latin American ecofeminism emerged along with feminist theology out of the theology of liberation movement (Ress 2006, 30). However, these movements have parted ways. Liberation theology started out as a movement that put the poor at the center of their discourse. It nevertheless laid the foundation for feminist theology and ecofeminism. When these movements began articulating their own discourse, the topics moved toward the inclusion of women and nature as victims of an oppression exercised by androcentric, patriarchal, and capitalist entities. Ecofeminists argue that nonhuman victims of exploitation such as ecosystems are affected by practices of destruction that mirror oppressive activity. The poor, indigenous, blacks, women, and nature or planet Earth are all affected by oppression.

The task of ecofeminism is described by Rocheleau, Thomas-Slayter, and Wangari (1996b, 9), 'Feminist political ecology addresses the convergence of gender, science and environment in academic and political discourse as well as in everyday life and in the social movements that have brought new focus to this issue'. For instance, ecological activism movements in Latin America consist mainly of women. Moreover, the higher the members of the movements climb up in social hierarchy and ascend the economic ladder, the fewer the women are among them. 'Historically, women's social contributions have been recognized by the movement's leaders, but women have remained mostly in the background of the economic and political arenas' (Rocheleau et al. 1996a, 28). With the socio-economical conditions in Latin America as they are, compounded by the effects of climate change and ecological devastation, the most oppressed group is that of women. As Aguilar notes, 'Although climate change will affect everyone worldwide, its impacts will be distributed differently between men and women as well as among regions, generations, age classes, income groups, and occupations. The poor, the majority of whom are women living in developing countries, will be disproportionately affected' (Aguilar 2009, 59). Women are in a predicament that renders them unable to emerge as agents of change. 'Climate change slows progress toward gender equality and thus impedes efforts to achieve wider goals like poverty reduction and sustainable development' (Aguilar 2009, 60).

Typically women act as the 'underbosses' of households and yet have no power in decision-making over their lives or their environment. 'Women and girls in developing countries are often the primary collectors, users, and managers of water' (Aguilar 2009, 60). Moreover, 'rural women in particular are responsible for half of the world's food production and produce 60–80% of the food in most developing countries' (Aguilar 2009, 60). Ecofeminism argues for the empowerment of women to mitigate climate change.[2]

Ethnic diversity offers multiple epistemological frameworks that need to be coordinated to cope with the climate crisis. The most committed activists in Latin American ecological movements come from the most marginalized social groups, the poor, women, and indigenous peoples. It is no coincidence that large numbers of activists belong to more than of these groups, because the worst affected group is that of poor indigenous women. The indigenous outlooks provide a framework that is not hegemonic: in this epistemology, nature is not a separate ontological entity, but instead part of the ontological realm shared also by humans. These outlooks have been systematically disempowered and marginalized by the Western framework.

## 4. Indigenous ecopragmatism

Various Latin American thinkers consider a synthetic approach to the climate crisis. One example is Rodriguez's 'ecological pragmatism'. He aims to recapture differentiated epistemological and rational systems of rural and indigenous communities for a critical incorporation into the environmentalist movement. He argues for the inclusion of these groups into the discussions and decision-making procedures. This approach offers a viable alternative against the effects of colonialism.

When indigenous groups in the Americas were forced into foreign European hegemonies, this corrupted their respective cultural identities and knowledge. The destruction of tropical forests for the extraction of its resources not only reduces biodiversity, but also threatens indigenous populations. These communities are consequently displaced, their social cohesiveness is shattered, and typically they end up being plagued by disease, starvation, and utter misery.

Official Latin American policies concur in the need to mitigate climate change. They also rely on the same growth-based economic model that primarily serves the interests of the developed countries. Such policies wrongly assume that the industrial way of life is not in its entirety responsible for its climatic consequences, while keeping alive the notion that underdeveloped countries ought to aim at growth through alternate methods while keeping the growth model largely intact. This creates a divide between the politics of the developing countries and their citizens and inhabitants. In the global south, and especially in Latin America, policies replicate the *status quo* of the North, while the masses grow ever more suspicious of their governments.

It seems fairly obvious to state that policies, from developed and developing countries, have failed to deal with climate change. This failure of policies makes one turn to the people, and their quotidian practices and interactions with their environment. Of particular relevance here are the outlooks of Latin American agents who have historically been ignored – people with indigenous and thus alternative lifestyles that foment sustainability on all biophysical levels.

In Ecuador and Bolivia, one such alternative framework is gaining force with the Quechuan concept of the *Sumak Kawsay*, which literally translates to 'good living'. At the state level these two nations succeeded to include all their ethnic groups. Guided by the notion of good living, both countries have also adopted the inclusion of nature in their constitutional framework of rights.

The Andean notion of good living may appear vague and simplistic. And it may be difficult for *Sumak Kawsay* to be made consistent with the neoliberal paradigm. The concept entails the notion of humans living in harmony with their surroundings. It sets itself apart from the dependence of economic growth, commoditization of nature, and accumulation of capital. Good living, in the Quechua sense, serves as a cultural alternative to the neoliberal practices and their pernicious impact on Latin America.

Historically the concept is rooted in Quechua tradition. The constitutional language of Ecuador and Bolivia showcases the historic-mythological roots of the indigenous traditions of the Andes. This alternative worldview survived the systematic marginalization by neoliberalism. Since the wars of independence in Latin America, the new nations pursued political agendas to make non-European groups invisible in order to present their own national identities as cultural extensions of Europe.

The constitutional language is mostly written in Spanish but includes the Quechua concepts *Sumak Kawsay* and *Pacha Mama* (Nature). *Sumak Kawsay* is the principle of the interdependence of living and in animate beings in nature; 'being-in-the-world' means to establish and maintain harmonious relationships with other forms of life and the environment as a whole. It implies a sense of respect toward nature. The principle recognizes the value of both nature and humans without drawing an ontological distinction between them – humans are 'with' and 'in' nature. The goal of the pursuit of harmonious relationships is equated with good living.

Western culture has defined good living in a radically different sense. Success in life is not measured in overall well-being. Rather, living well is associated with the accumulation of wealth. Thus, the western ethos revolves around consumerism – 'a cultural pattern that leads people to find meaning, contentment, and acceptance primarily through the consumption of goods and services' (Assadourian 2010, 8). Many Latin American thinkers blame the climate crisis on Western-style consumerism and its economic model.

The indigenous approach leads to a reappraisal of the concept of nature. The Western conception of nature as an instrumental entity is replaced with a conception of nature as a relational whole (Cormac 2010, 143). Only recently this approach is being taken seriously in the West as the 'new' framework for confronting the environmental crisis.

For indigenous cultures, the principle of good living has practical consequences. It is not some sort of abstract imperative. On the contrary, it enables concrete and commonsensical modes of sustainable behaviors. One consequence is the normatively embedded practice of *using only what is necessary*. Another is the consequence of *living with the least possible means.*

There are, of course, a few shortcomings about the principle of good living. There is the danger of making these groups responsible for finding a miraculous solution to a crisis that they did not create. Then there is the difficulty of implementing such indigenous practices in the developed nations. The British climate ethicist James Garvey writes, 'perhaps there are some calls for a new ethic which we simply cannot heed. Or maybe we can only aim to heed them someday, perhaps work our way towards them and hope for wisdom'. Yet we may wonder whether we can afford waiting much longer. Wisdom, it seems, is not a luxury anymore. What is needed, it seems, is a philosophy that contains not only an ethical framework to engage with the climate crisis on normative terms but also an ontological framework to ground a transformation of values from the conventional to the indigenous.

Latin American thought has established just this groundwork. As the Argentinean environmental thinker Barra Ruatta points out, Latin American (ecological) philosophy has established itself as being 'original and authentic'. Its originality and authenticity rest on the conceptual approaches of indigenous and ecofeminist groups that integrate social and ecological injustices in the same ontological and epistemological frameworks, which is something that the current Western paradigm fails to accomplish. This new Latin paradigm is an outlook that contains different worldviews, such as the indigenous, feminists, or theologians, while holding a sensitive and holistic notion of nature. This notion is the result of a shift from anthropocentrism to a cosmocentrist understanding of reality, and it identifies the essence of nature with the complex interwovenness of things and beings.

Most indigenous populations, such as the Ecuadorian Quechua and the Colombian tribes in the Sierra Nevada de Santa Marta, have preserved such cosmocentrist perspectives in their cultural practices; to them, there is nothing 'green' about it. It is only when seen in the Western optics of environmentalism that these populations appear so. As the Colombian anthropologist Ulloa points out, indigenous groups have entered the international stage with the acquisition of 'eco-identities' that consolidates them as a group distinct from Western modernity. This process of greening their public identities began during the commemoration of the 500th anniversary of the European discovery of the New World. There, indigenous groups established sharp differences, based on a fundamentally different notion of nature. They do neither look at nature as a mere 'basket of goods' nor consider humans to have a god-given right to exploit its contents. Their outlook, of a 'harmonic relationship with nature', originates in an 'ancestral knowledge that has allowed the continuity of biological diversity' (Ulloa 2005, 166).

## 5. Concluding remarks

For achieving a sustainable civilization, humanity must pool its efforts, and not only of the technical, scientific, and legal kind. Cultural and esthetic values must be reappraised. Sustainability is only feasible with whole global community. A power shift is needed for other cultures to be heard and to contribute both to theory and to practice. In this way, true progress of humanity can be attained.

Although philosophy has historically been at the forefront of inquiry, it has woefully fallen behind in the treatment of climate change. This 'retardation' reveals a shortcoming of the Western paradigm. Latin American thinkers have not only the necessary elements to formalize a critical approach to the climate crisis, but also the duty to do so. In this way the climate crisis is an opportunity for Latin American philosophy to seek a new enlightenment to end the overall existential crisis of the region.

Recent decades have seen a great upsurge of philosophical productivity in Latin America. The philosophy of liberation, particularly the work of Argentine-Mexican philosohper

Enrique Dussel, seeks to form a critique against Eurocentrism while advocating a new way of thinking based on the difference created by colonialism. Relevant is also the research of the Argentine semiotician Walter Mignolo on colonialism and the geopolitics of knowledge. To date, however, these new ways of thinking have touched on the climate crisis only in passing. But they do offer critical insights, such as the power relations that derive from colonialism, that are relevant to develop the foundations of a Latin American philosophy of climate.

Latin America is one of the global regions most affected by the climate crisis. Its forests have immensely suffered; its natural resources have been severely depleted; its glaciers are disappearing faster than elsewhere; its rain forest in the Amazon is on the verge of drying out to a savannah. Latin Americans need to leave this path of destruction. The philosophical culture of Latin America offers the prospect that this can be done.

## Acknowledgements

The author thanks Ivan Marquez (Bentley College) and Adriana Novoa (University of South Florida) for helpful suggestions to earlier drafts of this paper, and Martin Schönfeld (University of South Florida) for stylistic editing of the text.

## Notes

1. For a detailed analysis, see Arturo Escobar's 2010 essay, 'Ecologías Políticas Postconstructivistas'; e.g. http://www.unc.edu/~aescobar/text/esp/escobar.2010.EcologiasPoliticasPostconstructivistas.pdf (accessed February 1, 2011) or http://www.unc.edu/∼aescobar/text/esp/escobar.2010.EcologiasPoliticas Postconstructivistas.pdf (accessed February 1, 2011).
2. The word 'Empowerment' does not directly translate to Spanish. Mariano Velázquez de La Cardena's Velázquez Spanish and English Dictionary (El Monte, Cal.: Velázquez, 2003) offers the word 'Autorizar' as first translation, which is not a good semantic match, as it suggests a permission granted.

## References

Aguilar, Lorena. 2009. Women and climate change. In *State of the world 2009: Into a warming world*, ed. Linda Starke, 59–62, A Worldwatch Institute Report on Progress toward a Sustainable Society. New York/London: Norton.

Assadourian, Erik. 2010. The rise and fall of consumer cultures. In *State of the world 2010: transforming cultures – From consumerism to sustainability*, ed. Linda Starke and Lisa Mastny, 3–20, A Worldwatch Institute Report on Progress Toward a Sustainable Society. New York/London: Norton.

Boff, Leonardo. 1995. *Ecology and liberation: A new paradigm*. Mary Knoll, NY: Orbis.

Cova, Nancy Santana 2005. *El Ecofeminismo Latinoamericano, Las Mujeres y la Naturaleza como Simbolos*. Cifra Nueva, Universidad de los Andes-Trujillo.

Cullian, Cormac. 2010. Earth jurisprudence: From colonialization to participation. In *State of the world 2010: Transforming cultures – From consumerism to sustainability*, ed. Linda Starke and Lisa Mastny, 143–8, A Worldwatch Institute Report on Progress Toward a Sustainable Society. New York/London: Norton.

Driessen, Paul. 2003. *Eco-imperialism: Green power, Black Death*. Washington: Free Enterprise.

Escobar, Arturo. 1999. After nature, steps to an antiessentialist political ecology. *Current Anthropology* 40: 1–30.

Garvey, James. 2008. *The ethics of climate change: Right and wrong in a warming world*. London: Continuum.

Gudynas, Eduardo. 1992. Los múltiples verdes del ambientalismo latinoamericano. *Nueva Sociedad* 122: 104–15.

Gudynas, Eduardo. 2009. Climate change and capitalism's ecological fix in Latin America. *Critical Currents* 6: 36–41.

Leff, Enrique. 1994. El movimiento ambiental y las perspectivas de la democracia en America Latina. In *Retos para el desarrollo y la democracia: movimientos ambientales en América Latina y Europa*. Venezuela: Nueva Sociedad.

Portillo, María J. Saldaña. 2003. *The revolutionary imagination in the Americas and the age of development*. Durham/London: Duke University Press.

Ress, Mary Judith. 2006. *Ecofeminism in Latin America*. Mary Knoll, NY: Orbis.

Rocheleau, Dianne, Barbara Thomas-Slayter, and Esther Wangari, eds. 1996a. *Feminist political ecology: Global issues and local experience*. London/New York: Routledge.

Rocheleau, Dianne, Barbara Thomas-Slayter, and Esther Wangari. 1996b. Gender and environment: A feminist political ecology perspective. In *Feminist political ecology: Global issues and local experience*, ed. D. Rocheleau, B. Thomas-Slayter, and E. Wangari, 3–23. London/New York: Routledge.

Ruatta, Abelardo Barra. 1996. *Antiecología. Apuntes de una filosofía y paradigma ecológico*. Buenos Aires: Espacio Editorial.

Ruatta, Abelardo Barra. 1996. *Antiecología. Apuntes de una filosofía y paradigma ecológico*. Buenos Aires: Espacio Editorial.

Ulloa, Astrid. 2005. *La construcción del nativo ecológico*. Bogotá: Instituto Colombiano de Antropología e Historia.

# Watsuji Tetsuro, *Fudo*, and climate change

Bruce B. Janz

*Department of Philosophy, University of Central Florida, Orlando, FL, USA*

In this paper, I wish to consider Watsuji Tetsuro's (1889–1960) concept of climate (*fudo*), and consider whether it contributes anything to the relationship between climate change and ethics. I will argue that superficially it seems that *fudo* tells us little about the ethics of climate change, but if considered more carefully, and through the lens of thinkers such as Deleuze and Heidegger, there is ethical insight in Watsuji's approach. Watsuji's major work in ethics, *Rinrigaku*, provides concepts such as between-ness and trust that enable his philosophy of climate to move from a theory of national characters (as *Fudo* is often seen to be) to an approach to living well within one's milieu.

## 1. On the possibility of an ethics of climate

The question of the human relationship to climate seems to be an irreducibly moral question (along with, of course, a scientific one), but formulating an ethics of climate may not be as straightforward as it seems. This is the case because of the history of ethics in the West, and because of the unique nature of climate as an object and a context of ethical reflection. In this paper, I wish to consider Watsuji Tetsuro's concept of climate (*fudo*), and consider whether it contributes anything to the relationship between climate change and ethics. I will argue that superficially it seems that *fudo* tells us little about the ethics of climate change, but if considered more carefully, and through the lens of thinkers such as Deleuze and Heidegger, there is ethical insight in Watsuji's approach.

The history of ethics in the West is sometimes seen as an ever-expanding circle. That is, the agents of ethical action, and the potential objects of ethical concern, are often seen as having broadened over time to include a wider range of actors and objects. Ethical agents must have will, which limits agency to humans, and more than that, humans with the full capacity of will, and a reasonable access to knowledge and ability. The objects of ethical concern, likewise, historically have been those which could also be agents, or at least potentially agents. So, for most of the history of ethical theory in the West, ethics has been about how humans act toward and among other humans, and in fact, true moral actors.

It is because of this that environmental ethics has a recent history. The central meta-ethical question in environmental ethics has been, Can non-humans be objects of ethical concern? If so, why, and on what grounds, and to what extent? If not, why not? The anti-vivisectionist movement in nineteenth century England was predicated on the extension of ethical concern to individual animals, on the grounds that they could feel pain. In the US, the ethical discourse on wilderness (John Muir) and later on biospheres (Aldo Leopold) added to the list of things which may be objects of ethical concern, and did so in part because of an indirect sense of concern for humans (i.e. the environment that humans would inherit in the future), and in part because of a direct sense that an ecosystem can be an ethical object apart from the question of whether its individuals feel pain. Leopold, for instance, was clear that ethics at times might require a culling of herds, to bring an ecosystem back into balance. Individual pain, therefore, was not the central issue in his environmental ethics, but balance and harmony was. One could see the cause of the dis-harmony in nature as the result of dis-harmony among humans (Bookchin's social ecology makes this move, as do some strains of eco-feminism). Or, one might take a larger-scale approach (e.g. Lovelock's Gaia hypothesis, Arne Naess' deep ecology) anthropomorphizing nature in a quasi-Romantic manner in order to restore its ethical standing in relation to humans. All of these are directed at addressing and rectifying the central tension in the history of ethical theory, which is fundamentally about the relation of humans to each other. It is important to note that I qualified this description at the beginning as 'Western' because there are clearly different approaches to ethics in traditional and aboriginal systems, among others. In other words, the issue of the expanding circle is not necessarily the historical trajectory of discussions of ethics in other places.

I give this simplified overview in order to raise the question of where climate ethics might fit. On the surface, it seems as if it must be a type of environmental ethics. And yet, if we see environmental ethics described as an expanding circle, it is difficult to see how climate might fit in. There is, of course, the anthropocentrism that appeared in earlier forms of ethics, which would impute value to climate based on concern for present and future humans. But there is even less likely to be an object of ethical concern with climate than there is with the ecosystem. If we consider climate as a more dynamic and rarefied object within the group of environmental objects that have become part of ethical discourse due to the expanding circle, it is hard to see how one of two particular kinds of anthropocentrism can be avoided. The first version of anthropocentrism in question here is that which says that nature in general and climate in particular is an object of ethical consideration because it affects humans, and the second version is that which assumes that nature or climate is ethically significant intrinsically, since it takes on the characteristics of humanity (the Romantic approach mentioned earlier). In both cases, nature is significant because of us, and both cases are susceptible to dismissal, the first because one might think of a technological fix that breaks the causal link, and the second because the characteristics of humanity seen in nature are almost always asserted and assumed rather than defended.

But we might be able to think anew about climate if we take it out of the expanding circle metaphor that is part of so much of the discourse of environmental ethics. It is certainly tempting to think of climate ethics as a form of environmental ethics, due to the fact that climate is the environment and also affects other elements of the environment. But it is also possible to think of climate ethics not as a form of environmental ethics, but rather, to conceive of climate as a different object of reflection. Or, put more carefully, perhaps not an object of ethical reflection, but a context of ethical reflection, a place that frames ethics itself. The first thing to do will be to consider the nature of climate.

## 2. In the middle of climate

Climate is not weather. If weather is the interaction of physical features of the world, climate is the relationship between those forces and biological beings, in particular beings who are capable of having meaningful relationships with the world. Climate matters to someone, or to some species or ecosystem, and if there is no possible way of mattering (for example, with atmospheric phenomenon on other, presumably lifeless planets), we do not yet have climate. Climate affords ways of being.

Put another way, we could consider weather to be temporary phenomena, which we approach in a cause-and-effect manner (and indeed, humans are implicated in that cause-and-effect matrix, in that both our conscious and unconscious actions are influenced by weather), while climate is a longer-term phenomenon, which brings human ways of being into reality. It is one thing to put on a coat in cold weather, and quite another for an entire people to structure their cities, buildings, schedules, interactions and habits around predictable, long-term climactic conditions.

Climate is milieu, that which we find ourselves in the middle of. 'Ourselves' does not only refer to us in the limited sense of me and beings of my species, but it refers to the larger scope of beings to whom the milieu might matter. Those beings, as do we, intervene and inscribe the world in response to climate, and those inscriptions then become part of the milieu. The beaver would not build a dam were it not for its adaptive and creative inheritance from millions of years of evolution, which not only gives it protection, but alters the land, making it legible to beavers and other species, but also ultimately influences the climate by changing water flow patterns. We, too, inscribe the land in response to climate, thus creating climate through our myriad actions. Those milieu-relations are not all the same for all species – meaning may differ quite radically between species, even those involved in the same ecosystem or interdependent in various ways. We might differentiate between various kinds of symbolic systems of meaning, or alternately, a phenomenological/hermeneutical version of meaning that requires something like intentionality or self-consciousness. For our purposes, though, the line between weather and climate will be between sets of relations that we regard as being cause-and-effect, and which are best analyzed through the sciences, and sets of relations we see as meaningful, and are better analyzed through a combination of science, social science, and humanities.

We do not only create climate, though. Watsuji Tetsuro (1961, 2) begins his book *Climate and culture* (*Fūdo ningen-gakutedi kōsatsu*, or *Fudo*) by pointing out that our knowledge of climate is irreducibly experiential. 'How can we know the independent existence of the cold before we feel cold? It is impossible'. He might here be thinking of weather when he gives this example, but the accretion of experience, and the cultural production of representation about that experience, and which assumes particular kinds of climactic experience, is what makes up the milieu. So, not only is climate equivalent to milieu, the place which matters to us, but that mattering is mediated in a bodily fashion, at least at its most basic level. It is not, in other words, a 'higher level' structure of meaning such as narrative or myth, but something much more immediate. This is not to say that cultural narrative cannot come into the question when thinking about climate (living in Florida now has shown me that what people call 'cold' is vastly different from what was cold in my native Canada), but that meaning begins with experience in a very direct manner.

All of which raises questions if we try to use Watsuji within the context of recent debates on climate change. The use of the term 'climate' has seemingly shifted, toward a focus on the natural world. It still has human activity as its central problem, but that problem has largely been redefined as one of cause-and-effect mechanisms. The debate revolves around the question of whether or not there is climate change, and whether or not it was caused by human activity (and, assuming it was, which human activity caused the effects in question, and how can new human activity have ameliorating or healing effects). The question of the milieu, that is, the world we find ourselves in the middle of, has significance in the contemporary debate only in the sense that climate is the condition for the possibility of human existence. It does not tend to have phenomenological significance in the debate.

A second reason why using Watsuji to analyze the contemporary debate on climate raises questions is that the experiential underpinning of Watsuji's phenomenology does not fit well with the current debate. The changes in global climate are made available and justified through the use of scientific technology, which sometimes is not directly substantiated by embodied experience, and indeed can sometimes seem to be contradicted by it. It is already a common joke to point out that 'last winter was cold, and so global warming must not be occurring'. Of course, the temperature in a given place at a given time says nothing about global trends, but if embodied experience cannot be a reliable indicator of global climate change, Watsuji's link between climate and bodily experience seems tenuous.

And yet, I wish to think about current discourse on climate using Watsuji as a guide. His apparent inappropriateness to the occasion is not what it seems. In fact, I will argue that there must be more than just good science in addressing the questions raised by climate change. There is also the need to think about why climate should matter, and why it should matter as more than just the condition for the possibility of human existence.

## 3. *Fudo* as milieu

So, let us see whether Watsuji gives us some insight on how to address a situation that he did not imagine when he wrote *Fudo* in 1935. Watsuji Tetsuro recognizes that the world is an always already meaningful place for us. Part of his inspiration is Heidegger, and his conviction is that Heidegger's focus on time told only part of the story. Watsuji sees place and space as just as determinative of human consciousness and culture as time. In this, he prefigures Heidegger's move to a more explicit thematizing of place, later in his career (Malpas 2006). For Watsuji, *fudo* becomes a way of differentiating several different national characters. His three basic climate types, monsoon, desert, and meadow, correspond to different world cultures. The monsoon type includes Japan, China, India, and the rest of the coastal belt of Asia. The desert type refers to Arabia, Africa, and Mongolia. The meadow type mainly refers to Europe.

Some thinkers inspired by Watsuji took this taxonomy to suggest a geographical determinism. Somehow, climate was supposed to be the cause of individual and collective character. Watsuji is clear that he does not intend this sort of determinism (although, it is an open question as to whether he is nevertheless committed to it). What he does intend is to use geography as different spaces in which various values can emerge (Nagami 1981, 282; Berque 2004, 391). In this, Watsuji emphasizes the interrelatedness of place and social/cultural action. His concern is to resist what he believes he finds in Heidegger, and most other Western thinkers, which is a tendency toward individualism. The move to place is meant as a counterpoint to Heidegger's preoccupation with time in his early philosophy, which Watsuji sees as privileging individual intentionality. While this reading of Heidegger may not be fair (it was the standard way to take him for a long time, a way that read him primarily as an existentialist, but which cannot be sustained with a careful understanding of his work), and while this reading, even if it is correct, cannot apply to the later Heidegger, it was Watsuji's reading, and it sets up the project of restoring a place for collective or social intentionality and character.

There are problems and questions with Watsuji's move to introduce place as an antidote to the individualism that he sees the focus on time as bringing. For one thing, his version of climate only refers to human existence. Climate becomes milieu, but the question of value shifts, at best, from the individual to the social group. The environment does not have a direct value, but is only of interest inasmuch as its uniqueness and variability brings about different kinds of collective human existence.

Second, his version of climate clearly has an agenda. Japan comes out on top. This should be no surprise – Heidegger also can be read in many ways as having Germany come out at the pinnacle of civilization, and he is by no means the first to have this ethnocentricity. But it raises the question: What would a sense of climate look like if it was not ethnocentric, and if it was not Whiggish?

Third, there is a sense of defending nationalism and statism here. Watsuji naturalizes cultural forms, making them quite conservative. What is, must be the natural result of the place in which it is developed. That means that any radical social change cannot happen, or at least, cannot be defended.

So, there are issues if our task is to analyze Watsuji's argument. His version of Heidegger could be questioned, and his solution seems to naturalize social development. But the primary point here is not to analyze Watsuji's argument, but to see what he might contribute to the contemporary climate change discussion. What is most interesting about Watsuji's approach to climate is the relationship between climate and forms of life. He tends to focus on cultural forms and habitus, as he described the different climactic areas around the world. The central question that Watsuji gives us access to is this: given that climate change is the change of milieus, and given that milieus are necessary for differentiated human cultural development, what happens to cultures and to thought if all climates undergo radical and sudden change?

This may seem like a strange question. The usual question, and for most people the much more urgent question, is whether climate change will destroy all human life, or render it unrecognizable by changing our patterns of habitation and travel irrevocably. It may seem odd to ask about how cultural milieus might change due to climate change. And yet, this seemingly inappropriate question will, I believe, open the door to some useful elements of the climate change discussion that have not yet been addressed in any systematic manner.

Those who have tended to minimize the effects of climate change have offered several arguments as to why it will not be something worth worrying about. First, some have argued that climate change is not really the fault of humans, and it is a natural cycle that the world goes through. This, of course, does not deny that we may be facing cataclysmic climate change, only that there is little or nothing we can do about it, since we did not cause it in the first place. Second, others have argued that we can adapt relatively easily to climate change. For every place that turns into a desert, another will turn into a temperate place, supporting human life where conditions were previously hostile. Third, some argue that human technological ingenuity will address any problem that human technology has caused.

If put into the terms of Watsuji's concept of climate, the first of these arguments suggests that the shift in climate is something that cultures will simply have to adapt to. It is part of the inevitable nature of the world. The problem, of course, is that the argument is based on speculation, and has to ignore a large amount of established science. But it does point to one thing – that whether climate change is due to human causes or natural cycles, a change in climate will require a change in culture. At least, it will require such a change in Watsuji's world. We shall see if this still applies today.

The second argument assumes a relatively simple transformation of cultures to new characters. Even if we do not assume that Watsuji is a geographical determinist, his view of the relationship between culture and place would not allow such an easy move into something radically different.

The third argument is in some ways the most widespread, and the most interesting. Heidegger established that our relationship to technology is such that we do not simply use technology with no implications for ourselves. Technological optimists tend to think that new tools and techniques merely add to our abilities, without changing us in any way. I am the same person, so the belief goes, but with this new gadget that gives me some new capability. 'The Question Concerning Technology' was eloquent in pointing out the limits in such thinking. Our technology turns us into users of that technology, and when that technology depends on turning the natural elements of the world into standing reserve, the result is that we come to think of ourselves as outside of the world, all-powerful and able to manipulate the world into whatever form we desire. The world becomes plastic, while we also become plastic. Heidegger draws the distinction between what is 'correct' and what is 'true' here – while it may be correct to produce a new piece of technology to solve a problem, the truth is that we cover over the multiple possibilities of the world, while also turning ourselves into nothing but tool-users.

The technological dream which underlies the hope that the climate crisis will be solved by a new piece of technology ignores the fact that we are just continuing a relationship to technology that produced the problem in the first place. We believed that we would be able to turn the world into raw material for our uses. In fact, we could not, and the world is now answering back. At the same time, we continue in the belief that more mastery over the world will solve everything.

This is perhaps where Watsuji becomes most interesting. He is trying to establish what I have elsewhere called a 'philosophy-in-place', that is, the recognition that concepts come from places, they are 'owing', or have 'debts and duties', to use a phrase of Derrida's. Our sense that we have mastery over the world takes us out of our milieu, out of climate in Watsuji's sense of the term. Even if the argument is true, that we can find some technology to 'deal with' changes in climate, we cannot deal with Watsuji's sense of climate. The climate-solving technologies, just like the climate-changing technologies, turn us into beings and cultures removed from our climate.

In this, we face a more serious loss of place than even globalization portends. Globalization tends to undermine attachment to physical place by making those places homogenous with other places, by bringing goods, services, images, and knowledge into a new place that is shared with many other places. What is unique to a place can tend to get lost beneath the cultural force of these homogenizing elements. At the same time, though, globalized communication can bring people together in new interest groups. Where one may have previously felt alone in an interest or life experience, broadband internet and the proliferation of media outlets allows everyone to have a place, even if that place is virtual. Globalization, then, can be seen as the reconfiguration of place, holding forth some opportunities along with serious losses in other areas. It is, of course, a vastly more complex phenomenon than this suggests, but the point here is that there are opportunities for at least some. Whether those are worth the losses to physical place is, of course, another question.

When climate is lost, though, it is lost. There seems to be little or no compensation for the reconfiguration of human cultures, even if we imagine that climate change merely brings a reconfiguration of social experience, and not its complete destruction. Is there a 'philosophy-in-place' possible, when there is the adaptation to rapidly changing conditions?

In a sense, this is a very similar question to what occurs in cases of genocide, war, and mass exile or slavery. Whole cultural/intellectual milieus are changed, destroyed, or reconfigured in radical ways. What is maintained, is kept in memory, and hope, but not material reality. Even the process of healing can further the loss of the climate – witness the aftermath of the genocide of Tutsis in Rwanda. While they were slaughtered because of who they were, because of their identity, the frame of their existence now is one in which ethnicity is denied, in which all Rwandans are now just Rwandans, not Hutu or Tutsi. They are healed, but ironically their climate continues to be lost.

In other words, climate loss and climate change can occur in ways other than due to the biological and physical characteristics of the Earth. What is lost in climate change is not just some vague sense of 'our way of life', but our ecology of concepts and experiences, that make up the life-blood of our way of life. The threat, and therefore the stress, is very real under these conditions. In an important sense, climate change is about ecological philosophy, in this specific and idiosyncratic sense, even if it is not (as I suggested previously) about environmental philosophy.

The question of whether philosophy is possible in a world of climate change may seem like a minor issue, when faced with the possible destruction of human life. In fact, it is not. The philosophy of standing reserve, to use Heidegger's term, has become the implicit philosophy of a great number of people. It is the philosophy of the technological fix, the philosophy which supposes that there are no consequences to human consciousness for supposing that we can dominate and control the world. There is nothing new about this analysis – we can see it in Heidegger, but also any number of mid-twentieth century critics of technocratic society.

What is interesting to think about, though, is what we face now. We face human-caused climate change, due to our use of technology that has led us to suppose that we can control the world. Do we now use more technology to solve our problem, and does that lead us to further think that the world is malleable, shapeable by our agency, and designable into what we want? At one time, we thought that bodies were beyond our ability to mold; now, the dream of genetic engineering has us thinking that, in principle, disease and human limitation are things of the past. Careful biologists, of course, do not think in these terms. They are aware of the vast complexity of the biological organism, and the myriad ways in which organisms are embedded in their environments. If organisms could be controlled in a straightforward manner, we would already be reaping the fruits of having decoded the genome 10 years ago.

But public discourse moves more slowly, and still dreams of control and domination, both for biological individuals and for the climate in which they dwell. Watsuji would not be able to imagine climate as controllable; it is 'the agent by which human life is objectivised, and it is here that man comprehends himself; there is self-discovery in climate' (1961, 14). That is not possible if we simply control climate. Under those conditions it becomes our mirror, rather than that in which we come to self-comprehension. But we also live in the time that we do, in which decisions were made in the past to control all of human existence and all of nature. An ethics of nature will not be adequate if it only tells us the kind of ethical system to adopt if we are dealing with a blank slate. At the same time, continuing with a technocratic system of control will also not solve anything.

This is what we face. While it might seem as if Watsuji would have little to say to a situation like this, I would like to suggest that he is more relevant than it may seem. Reading *Climate and Culture* (*Fudo*), one gets the impression of a descriptive account that allows us to differentiate between cultures, and give reasons for why different forms of thought emerged. *Fudo* is irreducibly an account of place, and this is where we can begin.

Technocratic 'solutions' to global climate change tend to be universalist, even as they work from specific data. This universalism is a problem – it leads us to believe that, just as we have engaged in universal economic and technological activity to produce the problem, nothing less than a universalized effort will solve the problem. And, to some extent, this is true. Carbon reduction must be instituted, and that cannot only be the responsibility of a few. The reduction of energy use needs to apply around the world. And yet, those solutions cannot take root, since they come out of the same impetus that caused the problem in the first place. They may be the correct thing to do, but not the true thing (to use another of Heidegger's distinctions).

At the same time, solutions to climate change will necessarily be technological. They will just not be *solely* technological, nor will they come out of the same impulse as the approach to technology that caused the problems in the first place, an approach which is predicated on the concept of control. But what other option is there?

Bronislaw Szerszynski argues that '…nature is not a self-present equilibrium system; one therefore cannot simply represent the human metabolism with nature as somehow disrupting the stable metabolism of a pre-existing nature' (Szerszynski 2010, 14). Watsuji tends to think of climate as a stable entity, and so climate change is a disruption to a stable state. In this sense, Watsuji too fits into a technocratic mode of analysis of climate, despite his reliance on Heidegger. It is at this point that the direct usefulness of Watsuji's *Fudo* comes to an end. His version of climate does raise important questions about how our human way of being is related to the milieu in which we find ourselves. His characterization of that milieu, though, tends to be of a static or relatively stable place, one which is stable enough to engender particular large-scale cultural forms. Even if we see Watsuji's theory of fudo as less a theory of cultural essences and as more a theory of cultural attractors or central points, around which practices congregate and gain significance, we still have a theory that sees those cultures as largely self-contained and self-sustaining over time. Apart from the vast over-generalization that this represents, it also presents us with the problem of how to account for the fact that milieu itself continually shifts.

## 4. Detour through Deleuze

One possible direction to go at this point is with Deleuze. If we start with Watsuji's position, that there is a kind of philosophy-in-place which correlates with particular kinds of concepts and patterns of life, we can move to thinking about another kind of philosophy-in-place, one in which continual stressors and points of tension have given rise to concepts and patterns of life. At least one writer (Botz-Bornstein 2007) has seen a continuity between Watsuji and Deleuze, and at least one other writer (Jacobowitz 2005) has argued against there being a useful connection, at least at the level of accounting for culture, if not climate change.

To see how Deleuze might be of use, we need to go back to climate. Climate is milieu, but that milieu is filled with tensions and oppositions. It is not static, it is not just a mutually interlocking and supporting system of puzzle pieces. We find ourselves in this shifting milieu, but our relationship with it is not only that of figure and ground, or text and context. Classically, our ethics toward nature have been rooted in modernist systems based on either duty or consequences. We then either construct a system of rights, and try to determine what has those rights and what does not (and, how we can meaningfully speak of rights for entities that do not have autonomy), or we calculate consequences, and come up with a means of determining utility, harm, pleasure, pain, or some other proxy for a value that allows a justification of our actions.

As I argued elsewhere (Janz 2009), such systems tend to ignore place. They cannot easily attend to the fundamental difference and tension that exist in the world. They tend to look for commonality, so that a true calculus can be made, and a justification for action can be arrived at. What is missed is a form of knowledge that can discipline us, that can stand against our modernist tendency to control. In 'Thinking like a mountain', I argued that places 'know', in the sense that history is inscribed upon places, and that places can be 'read' in a variety of ways by the beings native to them. Neither ethics nor value are universal. It is located in a place, in the sense that ethical systems are worked out within a milieu, within a climate. Those milieus are not, of course, hermetically sealed off from each other, and so as different understandings of value are formulated, they must also encounter different systems. The tension between such systems, if allowed to continue and not be subsumed under a larger system rooted in a modernist fiction of universality, creates new values, new concepts that are adequate to a new kind of place.

Being disciplined by place-knowledge means recognizing the real multiplicity and diversity in nature, something that climate-change science in this best post-modern sensibility has a better chance of doing than does the modernist religion, economics, and politics that informs the alternative to this science. John Protevi, in a paper called 'Water' (Protevi 2007) which was part of a special issue of the journal *Rhizomes* on Deleuze and ecology, argues that the forms and processes of water give a great example of thinking in a Deleuzian manner, which is what I have been calling thinking in place. Protevi sketches in some detail the various cycles that water is involved in, and points out the forms of becoming that water is part of. It integrates into other cycles of many sorts, as it moves from its solid to liquid to gaseous form, and combines with other elements to form new cycles of moving, changing things. This is more than just 'ecosystem', a static term that attempts to freeze those processes for examination and analysis. Such an attempt is always an illusion, and this is something that climate scientists are well aware of.

Protevi moves from the analysis of the physical effects and becoming-cycles of water, to the politics that such cycles imply. We tend to think of politics as what comes after the analysis of the physical is complete, but Protevi argues that they are not at all separate. They are just more cycles, implied and brought into (virtual) existence by each other. We do not merely fight over water in an arid land – we change the land, we create objects, we organize institutions, we establish state mechanisms, and we establish our habitus in ways that are implied by natural cycles of the water.

In other words, the story we told earlier, about climate change as the domination of humans over nature is too simple a story. It does not acknowledge that human actions in the environment are one possible conclusion of life in that environment. They are not necessitated by it, certainly, but they were and are a set of choices in a milieu. They are continuous with it.

And, changing course will not come as a result of dominating the environment in a different manner. We will not find a new technology that makes everything right, any more than we will magically change peoples' attitudes so that climate change is reversed. The moral case for energy reduction in order to reverse climate change has little hope of success, because it relies on convincing a group of people who have other interests or priorities to abandon those interests, change who they are, and get on board.

Does this mean that all is futile, and we are doomed? No. It means that we must be disciplined by contingent nature, to see other 'lines of flight' (to use Deleuze's term), that is, to see other contingencies that can bring change. Some of this is happening already. Whereas once the case for energy reduction was made primarily on moral grounds, it can now be made on the grounds of economics and innovation. Some countries are seeing alternative energy technologies not as a second-best choice, but as an investment in a new set of possibilities (Stern 2009).

New concepts can be created in this environment. Climate change is the impetus for it. It is the condition of conflict that makes something new possible. The stress that people who are paying attention to the implications of climate science feel is interpreted as fear for survival, but that stress can also be the moment for something new.

## 5. Watsuji and the possibility of climate ethics

We find ourselves in a milieu, in 'climate', that governs what counts as a legitimate concept and what does not, what counts as a meaningful cultural form and what does not. Watsuji's static account of that climate, though, neither accords with the dynamic nature of culture, nor with the physical climate in which we exist, and furthermore, it is a resolutely human-focussed approach. Read superficially, Watsuji predisposes us toward the same kinds of patterns of thought that led to climate change in the first place. Deleuze allows us to make something creative out of the both the natural and cultural conflicts inherent in climate change. While technologies of control will just yield more unanticipated consequences, as they always have, there is the chance that a more self-critical science, one able to be productively critiqued by non-science in various forms, will allow us to live in a manner more integrated within natural cycles.

This version of Watsuji via Deleuze moves ethics away from a modernist Western conception of ethics, which tends to be about justification of action and the grounds for assigning praise and blame. It is, in some sense, the step before ethics, the frame for the world that makes an adequate ethics possible. A great deal of Western ethics begins from the premise that ethical attributes accrue to individuals or institutions like properties, and that the proper goal of ethics is to work out the best calculus (whether deontological or consequentialist in various ways) so that a desired goal is reached. It is clear that Watsuji does not have an ethics of this sort (nor, for that matter, does Deleuze). 'The locus of ethical problems lies not in the consciousness of the isolated individual, but precisely in the in-betweenness of person and person' (Carter, introduction to *Rinrigaku*, Tetsuro 1996, 10). For both Watsuji and Deleuze, the real question comes prior to ethics, and is epistemological – how do we conceive the world, and our place in it? What is our milieu?

Having said that, it is also true that for Watsuji at least, ethics is a central (perhaps the central) concern of his work, and *Fudo* is a work of ethics. The idea of in-betweenness is explicitly shown through interpersonal relations, but as with Martin Buber, it is possible to see a mode of betweenness that exists between person and non-person. In a sense, *fudo* is the context of ethical action, the web of relations and emergent properties that exist when we focus on relation rather than discrete elements of a system. For Watsuji, though, the betweenness that Buber develops does not appear at an individual level. Instead, the betweenness is the mediation of nature and nation (Nakashima 2002, 117). *Fudo* can be read as examining the natural and material conditions for a theory of collective nationalism. Watsuji's ethics, when he focusses on ethics alone in *Rinrigaku*, tends to not take up the earlier work in *Fudo* on the material conditions in which ethical action occurs (despite recognizing the importance of the climate in which ethical decisions are made, and criticizing Kant for being insufficiently sensitive to that climate – *Rinrigaku* 259), but is an examination of the basis of the ethical life, as rooted in a theory of the self. Even in *Rinrigaku*, Watsuji is less interested in providing a meta-ethics than he is in providing a framework for understanding how to orient the self in such a way as to lead to virtue.

*Rinrigaku* gives us plenty of description of the structure of human existence, including human spatiality, a concern first articulated in *Fudo*, but the goal of *Rinrigaku* is not to come up with an account of the conditions of national identities as it is to sketch out and ground an account of virtuous action. The term 'virtue' is used deliberately, and meant to evoke Aristotle's use of the term. Virtue is excellence, in Aristotle's sense, but an excellence that depends directly on climate, or on the interrelations of the milieu in which we find ourselves. The central concern, for Watsuji, and the thing that grounds ethics, is the nature and necessity of trust for ethics to be established, and the consequences of betrayal. Trust brings truthfulness, and therefore right action, while betrayal brings falseness, and therefore wrong action. Truthfulness and falseness are, therefore, as much moral categories as they are epistemological ones.

The payoff of truthfulness is more trust, and the result of betrayal is the lack of trust. Watsuji discusses the social implications of trust – the need to call on someone in time of crisis or need (*Rinrigaku* 266ff). Trust is, in fact, what makes the milieu (or climate, in Watsuji's sense) possible at all. The basis for trust is 'the rational will inherent in each person. To trust in a person is to place one's trust in the inner independence of this person' (*Rinrigaku* 268).

It would be too facile to say that the lesson to take from this is that we should all simply trust each other more. The real issue here is, why is there not trust, and what is the trust really placed in? How is trust an ethical virtue that is in some way corroded, both by the history and reality of climate change, and also the loss of Watsuji's sense of climate? How does the tragedy and crisis of climate change, established beyond a reasonable doubt by the best science available, become exacerbated by a milieu of lack of trust? And, is there a path back to a climate (in Watsuji's sense) where virtue can find its place?

For Watsuji, in trust 'one assumes in advance a decided attitude toward the future ... [A]dmitting that the future we aim at in the act of willing is, to begin with, tomorrow's betweenness, we must say that this betweenness will also arise in acting to do or undo something' (*Rinrigaku* 271). This future-oriented truthfulness comes connected to the attention to the milieu that we must have, that is, the recognition of the facts of the world, along with the creativity contained in the not-yet.

How does this help? Recall that this essay began by distancing climate philosophy from environmental philosophy, on the grounds that most environmental philosophy in the West is predicated on a version of ethics that begins with human to human relations and is extended by analogy in the 'expanding circle' model to include environmental objects. The alternative is an ethics of place, and Watsuji's concept of climate is really a kind of ethics of place.

Place ethics relies on something like a Deleuzian sense of creativity. In other words, ethics is not about proscriptions, primarily, but about finding ways for a true (that is, trust-bearing) approach to the environment, and to human social organization, can emerge. In a sense, the physical climate that has been the impetus for so much moral conflict is a red herring. Our addiction to oil is problematic whether or not there are climatic implications, because it is a technological solution that limits, rather than promotes creativity, and it undermines rather than supports the basis for truth-telling and trust in society. It is not simply that there are potentially dire consequences to not acting – the issues that have been correlated with climate change, such as species loss, habitat loss, and so forth are issues because 'Climate', in Watsuji's sense of the term, or human, cultural, and natural milieu, is made less diverse, less able to produce new concepts and new experiences. Climate change is only the most recent and most spectacular result of a way of dominating nature that has a long history to it.

The real lesson from Watsuji, though, is a positive one. If we think of ethics not as proscriptions, but as the question of the nature of the good, how to recognize it and achieve it, then Watsuji suggests something else. What is the good, in a world where climate change is a reality? The world is not a utopian one, in which we roll back the clock to an earlier, presumably idyllic age before we started using fossil fuels. There is no going back. There is, however, a new ability to recognize the limits of our previous ways of thinking and acting, and a new imperative to change. Of course, there are always ideological ways to dodge the imperative to change, and it may be that until something sudden and spectacularly bad happens, the ideology will continue to make itself felt and resist truth. As I said earlier, the point of ethics in the climate change discussion cannot be to convince those whose paychecks and investments depend on not believing.

For Watsuji, the good in relation to climate change is in the restoration and preservation of climate, in the sense of milieu. What will this accomplish? It will first keep us from well-meaning mistakes that can happen when we imagine a technological fix to an issue. Climate change does not just affect ecosystems, although it does that – it affects culture and thought. Failure to take seriously the subtle currents and flows of climate turns it into an abstraction, a thing that is not understandable. It becomes nothing but a tribal marker of politics and culture. People believe or they do not, not because of the science but because of everything else that comes along with believing.

The abstraction of climate change that dominates public discourse has no sensitivity to betweenness. There is no trust, and no truth. There is no flow, no dynamism, and no creation of anything new and vital. The climate crisis is surely a crisis of ecosystems and species, but it is also a crisis of discourse, imagination, and place. There is none of the subtlety inherent in a true knowledge of place, the kind of knowledge that allows other creatures to live and move and have their being. This, I believe, is the direction that Watsuji points when in *Rinrigaku* he speaks of the good (or the virtue) that truthfulness makes possible.

## References

Berque, Augustin. 2000. Offspring of Watsuji's theory of milieu (Fudo). *GeoJournal* 60: 389–96.

Botz-Bornstein, Thorsten. 2007. From community to time-space development: Comparing N. S. Trubetzkoy, Nishida Kitaro, and Watsuji Tetsuro. *Asian Philosophy* 17: 263–82.

Jacobowitz, Seth. 2005. Watsuji and Deleuze and Guattari in the climate of culture. CLCWeb 7:3 (September 2005). http://docs.lib.purdue.edu/clcweb/vol7/iss3/ (accessed May 8, 2009).

Janz, Bruce. 2009. Thinking like a mountain: Ethics and place as travelling concepts. In *New visions of nature: Complexity and authenticity*, Series: The International Library of Environmental, Agricultural and Food Ethics, ed. Martin Drenthen, Jozef Keulartz, and James Proctor, 181–95. New York: Springer.

Malpas, Jeff. 2006. *Heidegger's topology: Being, place, world*. Cambridge, MA: MIT Press.

Nagami, Isamu. 1981. The ontological foundation in Tetsurō Watsuji's philosophy: Kū and Human existence. *Philosophy East and West* 31, no. 3: 279–96.

Nakashima, K. 2002. Nationalizing nature: Discourses of fūdo and national environmentalism in modern Japan. *The Geographical Reports of Kanazawa University* 10: 115–25.

Protevi, John. 2007. Water. *Rhizomes Special Issue: Deleuze and Ecology* 15. http://www.rhizomes.net/issue15/protevi.html.

Stern, Nicholas. 2009. *A blueprint for a safer planet: How to manage climate change and create a new era of progress and prosperity*. London: The Bodley Head.

Szerszynski, Bronislaw. 2010. Reading and writing the weather: Climate technics and the moment of responsibility. *Theory Culture Society* 27, no. 2–3: 9–30.

Tetsuro, Watsuji. 1961. *Climate and culture*. Trans. Geoffrey Bownas. The Hokuseido Press. (Translation of *Fūdo ningengaku-tek kōsatsu*, 1935, 1943)

Tetsuro, Watsuji. 1996. *Watsuji Tetsuro's Rinrigaku*. Trans. Yamamoto Seisaku and Robert E. Carter. Albany: SUNY Press.

# Climate change and the ecological intelligence of Confucius

Shih-yu Kuo

*Institution of European and American Studies, Academica Sinica, Taipei, Republic of China*

Confucius is conventionally regarded as the founder of secular humanism and as a philosopher concerned about humans and culture. I would add to this that Confucius should also be read as an environmental philosopher. One reason is the pedagogical dimension in Confucianism, which points to Confucius as an environmental educator – not the least of which since much of environmental education relies on common sense and an enlightened collective self-interest. Another reason is an aspect I call 'ecological intelligence', which is a key feature of ancient Confucianism and a signpost for the civil evolution that the mitigation of climate change demands from global civilization today.

Environmental problems have steadily grown worse in recent decades and are now becoming systemic. Human-made climate change is symptomatic of this transformation from localized damage to global degradation. Climate change threatens to become the most serious environmental issue of this century. When we consider its causes, research is unanimous: we are too many and we want too much. We sacrifice nature for a more comfortable life. Can we change this? Is it possible to see ourselves and nature from another point of view? What would characterize a world in which can live a comfortable life, but also preserve the natural environment and sustain the Earth system? By a 'comfortable life', I mean a life that will not be looked back on with regret. What kind of ecological intelligence is needed to handle environmental problems, deal with climate change and maintain a habitable world?

Environmental ethics deals with how we should view nature and how we should act with regard to it. Since the 1970s, many theories have been proposed regarding these topics, but lately another approach has been developed: that of ecological intelligence. This approach involves not only ecological literacy but also acting wisely. What does ecological intelligence mean? What does it consist of? Is it a new concept, developed by current authors? Or can we trace its origin to a traditional school of thought? F. Capra in his *The Tao of physics* explores the parallels between modern physics and Eastern mysticism. By Eastern mysticism, he refers to the philosophies of Hinduism, Buddhism and Taoism. His exploration and discussion help us realize that modern sciences are not very far from Eastern mystical insights: 'There is an essential harmony between the spirit of Eastern wisdom and Western science' (Capra 1991, 25). If we accept his claim, we may be able to trace the concept of ecological intelligence back to Eastern schools of mysticism – one of them being traditional Chinese philosophy. Our question will be whether classical Chinese philosophy contains a point of view and a conception of nature relevant to environmental ethics in general and climate ethics in particular. To what extent, if any, did our ancestors possess ecological intelligence?

In this paper, I wish to argue, first, that the concept of ecological intelligence was anticipated and developed by Confucius; second, that his thought is based on the teachings of the *I-Ching*, especially the conception of nature in this classic; and third, that the teachings of Confucius can assist us in gaining the ecological intelligence that world civilization needs.

## 1. Idea of Tao (origin of Confucius' thought)

Though they conceive the concept of nature in the same metaphysical idealist terms, when it comes to philosophy of life, Lao-Tzu is a naturalist while Confucius is a humanist. Thus, while they develop the same view of nature, only Confucius displays ecological intelligence in a humanist context.

A few references can be found linking traditional Chinese thought and modern ecology. Usually, Taoism is considered to be more closely related to environmentalism with its idea of *wu-wei* (no action), which aims to leave natural processes undisturbed. Scholars also tend to separate Taoism from Confucianism and place them in opposition because of their differing applications of principles (Ames 1989). It might be true that Confucians favor regulating individual behavior within communities and create constraints according to rigid rituals, manners and proprieties. But this is derivative of ethical education, not of Confucius' cosmology or his concept of nature. My discussion will show that classic Chinese philosophy and environmentalist thought involve very similar views of nature and cosmos. F. Capra views classical Chinese philosophy as beginning in naturalism. The original purpose of the *I-Ching* (the Book of Changes) was divination based on physical phenomena observed in secular culture. When great intellects put those words to other uses, for example, to explain humanistic or spiritual matters, new interpretations of the text arise in later generations.

Western scholars – and not a few Chinese – tend to misunderstand classic Chinese philosophy and categorize it as a kind of religion (Tucker and Berthrong 1998, xxxix). Cheng gives an explicit elucidation of the concept of the trinity of Confucius, but he derived this concept from Christian theology (Cheng 1998, 225).

The origin in Confucius' metaphysics is from the *I-Ching* with the main concept of the Tao. When Lao-Tzu advocates doing nothing to nature and calls it doctrine of '*wu wei*', Confucius takes Tao as that which ancient intellects teach us, how we should behave, how to help nature, and why. We have one philosophical text and two interpretations of how to practice it. For Lao-Tzu's thought, we refer to *Tao-Te-Ching*. For Confucius, we refer mainly to his commentary on the *I-Ching* (易經繫辭傳) and other writings by his disciples and his grandson such as the *Doctrine of the Mean* (中庸) and *the Analects* (論語).[1]

The *I-Ching* is one of the oldest classics. It uses simple symbols to represent various natural phenomena and attaches a highly abstract meaning to each symbol. For more than 5000 years, various explanations and implications have been developed, and it is difficult to distinguish what real meaning is represented by the symbols. Some people use this book as means of divination, to foretell the fortune, or as a farmers' guide of agricultural planting. In academic circles, this book is the most important text to understand Chinese classic thought. It can be read as the first book of Chinese philosophy, though no author can be identified. In Confucius' commentary, it is supposed to have been written by Fu-si (伏犧氏) who is said to have discovered fire and taught people to plant crops.

There is an ultimate principle that runs throughout the book – the concept of the Tao, as an abstract name for a generative process. Following Tao, there are two streams that emerge: heaven and earth or *yang* and *yin*. These opposing phenomena are also symbolic of other ideas, including hot and cold, light and dark, moving and quiet, male and female. Following this, additional natural phenomena are generated and organized into eight trigrams. When two trigrams are combined, 64 hexagrams arise out of the eight trigrams.

Confucius considered the study of the *I-Ching* to be his life work, and wished to dedicate his later years to studying the *I-Ching* (Confucius 2010a, chap. 7, #17).

Confucius considers Tao metaphysically. In the *Analects*, he does not often discuss Tao because this work is not a philosophical treatise but a teaching guidebook, and he complains that although Tao is our main purpose and the moral principle underlying everyone's education, there are not many people who can reach it. Reading the *Doctrine of the Mean*, another book dealing with Confucius' philosophical thought, written by his disciple Zen-Zi (曾子) and his grandson Zi-Si (子思) but thought to be a reliable record of Confucius' views, we find that the original concept of Tao can neither be put in words nor even shown through people's actions; If you think you can reach it, it escapes you; It is the essence of our universe, a guide to our daily activities; It is ever moving.

Confucius thinks highly of the *I-Ching*. On multiple occasions, he mentions in his commentary that the *I-Ching* is the greatest and deepest book.[2] There are three ways the Chinese character 'I' (yì; 易) of the *I-Ching* can be read. One meaning is 'change'; another is 'easy' or 'simple'; and yet another is 'symbol' or 'symbolic'. The sense of 'I' (yì; 易) as change includes everything, and goes everywhere, far and near, without beginning or end. The Tao operates as a circular process; when anything reaches its an extremity, it reverses its course and cycles around to begin the process again. Such is the principle of change and reversal. The Tao of Qian (乾 or *yang*) is always in action. When it moves, it is strong and straightforward. The Tao of Kun (坤 or *yin*) is still. When it stands, it is quiet, smooth and pure. In their interplay, they are like the Sun and Moon, and the four seasons that help everything flow across each other in harmony. The essence of 'I' itself does not move, but has its two poles – Qian and Kun – between which there is interplay, and through which everything moves and grows. When yang and yin connect each other, the whole world starts. One variant sense of 'I' (yì; 易) as change to change by way of interplay between hard yang and soft yin. Six other trigrams give rise to specific natural phenomenon. Zhen (震) and Li (離) show thunder and lightning that touch things to bring them to life. Xun (巽) and Kan (坎) initiate the wind and rain to moisten and nourish the ground and crops. Dui (兌) and Gen (艮) create mountains and lakes. Li also represents Sun, and Kan, Moon, that move alternately as summer and winter in helping harvest and rest. Finally, the Tao of *yang* produces male and the Tao of *yin* produces female.[3]

Thus, Qian begins everything and Kun forms everything. The movement of their interplay is easy. Tao of Heaven works effortlessly; its method is simple; its action is easy to follow, and people abide it easily. This Tao is the ultimate principle of the universe. Once this principle and law are grasped, human beings become one of the three fates in nature. These three fates are Heaven, the Human, and Earth as the three elements of our concept of Tao. In the *Doctrine of the Mean*, Confucius shows how one can become part of these three, and how humans can cultivate virtue to join the trinity of the Tao. The sense of ease, the other sense of 'I' (yì; 易), points to another aspect of Tao: the principle of reversal. There are many examples of this in the *I-Ching*. For instance, the six lines of the first hexagram, Qian (爻), are traditionally interpreted as follows. The bottom line, called the First Nine, refers to a dragon still in hiding, meaning that one should not rush into action; the next line above mean that, it is time to act; it also brings hope and life to hard-working people; the third symbolizes a period of growing up; the fourth refers to the completion of education and to the departure from home; the fifth indicates the time to harvest; and in the last one, things end in exhaustion following a prosperous harvest, and the cycle begins anew. This is the natural law.[4] Confucius likens the ascent of the six lines of each hexagram to the discipline of human behavior. If this discipline is practiced in confidence, one can participate in the transformation and sustenance of Tao and form a trinity with Heaven and Earth.

Each of the 64 hexagrams consists of six lines which may be either broken (*yin*) or unbroken (*yang*). Each hexagram symbolizes the three categories of nature or the three elements of the Tao. The upper two lines symbolize Heaven, the lower two lines symbolize Earth, and the middle two lines symbolize Human. Their order cannot be mixed up, or else chaos appears. These three categories combined together transform the world. Thus, Human is in between Heaven and Earth. There is more than one way to explain the relationship among these three. In one sense, it means that Human, Heaven and Earth are equal and stand side by side. Human cannot dominate either Heaven or Earth. It also means Human cannot disturb either Heaven or Earth. In another sense, since Human is in the middle, it should be modest and always consult and respect both Heaven and Earth. However, if one wants to stand on equal footing with Heaven and Earth, one must first cultivate virtue. With this view, Confucius separates his ethical principles from those of Lao-Tzu, though they maintain the same view of nature.

Confucius's metaphysical thought concerns the actions of the Tao. Follow the Tao and everything comes naturally. Tao, Heaven, Earth and Human are equal and always interact with each other. Confucius, a humanist, emphasizes that 'only Human can promote Tao, not Tao promotes Human'.[5] The virtuous human being should aspire to the proper middle position of the trinity of the Tao. In the *Doctrine of the Mean*, he explains the virtuous behavior human beings should cultivate.

## 2. Confucius' view on nature

In this section, I will focus on Confucius' view, and use Lao-Tzu's thought as my frame of reference. As mentioned, the third sense of the Chinese character 'I' (yì; 易) of the *I-Ching* is 'symbol' or 'symbolic'. This aspect of 'I' suggests that human beings do not have any authority over nature and should not disrupt the process of heaven and earth. Confucius says that heaven never speaks, and yet the four seasons continue to change in time and generate all things.[6] Nature never expresses any opinion, and everything operates and follows order. There are many passages in his commentary on the *I-Ching* that express the same idea. For instance, all things under heaven follow nature, so there is no need for humans to worry. Summer and Winter appear in order (Confucius 1993, chap. 4). Heaven and Earth are like a pair of bellows: they are empty, and yet can never be exhausted. Do not say much and follow the middle way.[7] In classical Chinese philosophy, this cosmological doctrine is advocated by Lao-Tzu, Confucius, and Chuang-Tzu (Zhuangzi, 莊子).

Confucius applies principles of the *I-Ching* in his teaching and asks his disciples to cultivate virtue. A remarkable passage reflecting the Confucian view on nature is from the *Doctrine of the Mean*. In Chapter 20, when Confucius speaks to the Duke of Ai (哀公) about methods of governing, his words are short and few; they are elaborated by his grandson in an extensive commentary. As mentioned, there are three fates or elements of Tao, Heaven, Earth, and Human. The Tao of Human accords with the Tao of Heaven. The Tao of Heaven represents sincerity or truth (cheng, 誠). Making oneself sincere or truthful is accordingly the ethical path that follows the Tao of Human. When you are sincere, you can attain the Tao without effort or hard work. You will be like a sage with an easy manner.[8] In Chapter 22, Confucius points out, one can realize the human essence only through this sincerity or truthfulness. By realizing one's essence, one can also realize those of others. Fully realizing the essence of others extends to the essence of all things. Then one can assist Heaven and Earth in their transformation and sustenance.

Through assisting and shepherding the transformation and sustenance of Heaven and Earth, humans can evolve to be part of a trinity with Heaven and Earth.[9]

The Confucian conception of nature suggests that nature is never a mere resource to be used for human's benefit. When human beings utilize natural resources following the *I-Ching* to plant and harvest food, to cut trees for building shelter, they will hold a great respect for the natural world. This kind of ecological intelligence always shows in eastern traditional customs such as asking permission to hunt, thanking Heaven for a good harvest, etc. In Westerners' eyes, these customs look like some kind of mysticism. In the secular culture of the Chinese people, when farmers plant their fields, people plan their dates of marriages or funeral services, move in or out of a house, etc., they have to consult with the lunar calendar, also called a farmer's calendar, and its divination. If my interpretation of what Confucius says about the relation of humans and nature is correct, then Confucian thought already has what we need for today's environmental ethics.

## 3. Confucius' thought and environmental ethics

We are facing the challenges of global warming that affect everyone in the world. In the field of environmental ethics many theorists develop new approaches and provide advice to people. Wenz proposes one alternative view of environmental synergism (Wenz 2001). McCallum proposes ecological intelligence (McCallum 2008). I will argue that both thoughts have appeared in Confucius' thought in classic Chinese philosophy.

According to Wenz, there are three possible positions for valuing the natural environment: anthropocentrism, non-anthropocentrism, and what Wenz calls 'environmental synergism'. In an anthropocentric framework, environmental protection can be achieved by appealing to aesthetic criteria and by implementing the practice of deep economy. This practice, as McKibben points out, not only energizes local economies (by creating jobs in communities and increasing local profits) but also protects the environment, by discouraging the distribution and sale of goods produced far away, thereby reducing traffic intensity and attendant emissions (McKibben 2007). The non-anthropocentric framework involves individualistic and holistic variants. The latter includes approaches as diverse as animal liberation, ecofeminism, ethical theories based on the Gaia hypothesis, and Leopold's land ethics. Wenz argues that non-anthropocentric approaches in general fail to help us with adjudicating conflicts of interests between humans and nonhumans. As an alternative and third choice, he advocates environmental synergism. His definition is 'environmental synergists believe that synergy exists between respect for people and respect for nature. Overall and in the long run, simultaneous respect for people and nature improves outcomes for both' (Wenz 2001, 169). Wenz emphasizes the concept of simultaneity. It requires respect of nature and people simultaneously. He also says: 'respect for nature promotes respect for people, so the best way to serve people as a group is to care about nature for itself' (Wenz 2001, 169). An example would be to work with nature not by gaining unlimited power over nature but promoting biodiversity (Wenz 2001, 183). I believe that this approach might encounter some objections from conventional moral philosophers. For instance, a utilitarian would hold quantifiable considerations supposedly related to human happiness, such as material wealth, as the guideline for normative decisions, and not non-quantifiable, purely qualitative aspects of happiness, such as those related to respecting nature.

When we compare Wenz' idea of environmental synergy with the Confucian view of nature, we do not find much difference. Classical Chinese philosophers already suggest humanistic solutions to our environmental problems based on people's ecological intelligence. I will mention five points which coincide with Wenz' concept and also argue that the Chinese way of thinking is more convincing.

For classical Chinese thinkers, such as Confucius, the word 'Human' does not mean physical human beings but an abstract concept that indicates the sphere of humans in the cosmos. The Human is one-third of the concept of Tao that learns from nature. The other two are Heaven and Earth. Thus, Human is equivalent to Heaven or Earth, with the ability to learn from earth, heaven, Tao and nature. Confucius thinks Human, Heaven, and Earth form a trinity. Each one can complement the other, but only the Human element has the ability to praise and assist the other two in sustaining our future.

Being one part of the trinity of Tao does not mean that human beings will hold the greatest power to control and manipulate nature. On the contrary, as Confucius says in the *Doctrine of the Mean*, human beings have to be modest in front of nature. As mentioned above, only with the most sincere attitude can we realize the true character of the Human element. And finally, once we realize the essence of all things on Earth, we will be able to assist Heaven and Earth in their transformation and sustenance. Able to praise and assist Heaven and Earth in their transformation and sustenance, Human can form a trinity with Heaven and Earth. Sincerity is the beginning and end of all the other things. Without sincerity, there is nothing. The Human element is neither higher nor lower than all the other things. With sincerity, we are perfecting ourselves and also perfecting all the other things. Doing this, we need 'Jen' (仁) and 'wisdom' (智). The former means to love and respect from our heart and to let these virtuous characteristics become part of our virtue; the latter means to learn and be knowledgeable. Confucius recognizes that only the greatest intellects possess both.[10] Confucius pays great respect to these creators of the *I-Ching*. The Human should always be respectful toward ways of Heaven and Earth. This is the Human Way.

There are two aspects to look at in this trilogy. First, we pay respect to Heaven and Earth, and in return, Heaven and Earth help us by providing resources for Human. If we regard nature with love, respect and understanding, it is not possible that we will dominate and manipulate our natural world. Since we are sincere, we never cheat nature for our benefit. Whenever we utilize natural resources in helping us to produce living utensils, to rear livestock for food, or cultivate land, we have to follow sincerely what the *I-Ching* teaches in its divination. Farming according to dates on the lunar calendar has been practiced for thousands years in the eastern part of world. And to cultivate our virtue is one way that Heaven helps us. I believe this idea coincides with the concept of nature held by naturalists such as Foss today, namely, that nature holds no hostility and needs to be treated wisely (Foss 2009). As another classical philosopher, Hsun-Tzu (Xunzi, 荀子) said: 'Human is the noblest being in the world'.[11] This is the reason that Human can be part of the cosmic trinity and sustain Heaven and Earth.

In Confucius' view, there are not many people who can reach the level of the great intellects. We need these intellects to educate us. Teaching becomes one of the most important projects for them. Confucius famously claimed not to have written any original works but merely transmitted the wisdom of the ancient intellects instead.[12] He never claims to be one of the intellects. This way, he can be considered as an intellect by our moral standard.

Confucius can be regarded as a philosopher with ecological intelligence in the modern sense. I. McCallum in *Ecological intelligence* categorizes human beings as human-animals with human intelligence but with animal instincts, who act and react to their surroundings like animals do. A human being is a naturalist, a scientist and also a poet. Not only do we possess ecological literacy, but we also have to be involved in natural creation, with a poetic mindfield, as he puts it, caring for and respecting all creatures in nature. McCallum describes his own experiences of living in the wilderness to illustrate how he developed his ecological intelligence. If human beings can accept their limitation of being human-animal, and assist nature in its sustainability, the mindfield will be the place to launch our intelligence (McCallum

2008, 110–12). He does not provide any theoretical argument to justify his vision. We can ask why we should believe him, considering that he merely appeals to his personal experiences. I would think that individual intelligence alone is not enough we need everybody to realize, act coherently and follow the teaching of our sages and cultivate virtuous characteristics. I believe that we can strengthen McCallum's idea by appropriating Confucius' notion of nature and the Confucian view of the human-nature relationship.

Humans are not born with ecological intelligence; it has to taught and cultivated instead. As Confucius sees it, the best teacher of ecological intelligence would be the *I-Ching*, which he used as his own instruction manual. His written commentary helps to unlock the meaning of this ancient classic.

Contrasting the Confucian concept of ecological intelligence with the emerging reality of climate change evidently brings the absence or failure of such intelligence into view. It also puts climate change in a conceptual and ethical light. And it raises a question: how should one properly think of climate change in philosophical terms?

J. Garvey takes up this issue and discusses it in *The ethics of climate change*. When faced with the urgency of our climate crisis, doing nothing or even delaying action would not be the correct direction to take. He states that the longer we delay, the worse our future will be (Garvey 2008, 112). Doing something is a moral demand for action on climate change. Garvey's book is only one example of the many recent philosophical reflections on climate change. We need ecological intelligence and timely action. J. Lovelock shares a similar conviction, considering that human beings possess the intelligence to create the modern world, it is now time to use all our brains and energy so as to prevent Gaia as we know it from vanishing (Lovelock 2009).

## 4. Confucius' thought and environmental education

In his *Ecological literacy*, D. Orr reminds us that environmental problems are not a few people's interests but everyone's concern. He argues that all education is environmental education (Orr 1992). He identifies six basic building blocks of an educational system. They are: all education is environmental education; environmental issues are complex and cannot be understood through a single discipline or department; for inhabitants, education occurs in part as a dialogue with a place and has the characteristics of good conversation; the way education occurs is as important as its content; experience in the natural world is both an essential part of understanding the environment, and conducive to good thinking; and education relevant to the challenge of building a sustainable society will enhance the learner's competence with natural systems (Orr 1992, 90–2). When we review these building blocks and their implications for educational practice, we find that they comprise nearly all knowledge and techniques necessary for developing our ecological literacy. But in addition, we still need something that comes from our heart. That is to say, one needs to cultivate a feeling of care and an emotional bond toward one's environment in order to live wisely and respectfully.

Our empirical knowledge forms one-third of a framework of environmental education according to J.A. Palmer in his *Environmental education* (Palmer 1998, 269–72). Next to the empirical aspect, Palmer argues that environmental education also consists of an ethical and an aesthetic element. The three elements are not isolated from one another. Their overlap and convergence create a pedagogical synergy, the formative influence, which is the most important part of our education. It includes not only clear and profound knowledge of nature and environment, but also establishment of our attitude and skill in evaluation of our actions toward nature and appreciation for our natural world. However, Palmer does not develop his theory to the point of specifying what action we should take.

J. E. Foss maintains a more detached point of view in his philosophy of nature (Foss 2009). Nature is not perfect and does not possess any value at all. We cannot fantasize about nature in a romantic way. Nature is only a neutral ground on which we live. Foss criticizes some environmentalists who romanticize nature in quasi-anthropomorphic terms and regard environmental problems such as the climate crisis as nature's revenge. Foss suggests to conceive of humankind as the neurological center of our planet. He likens our planet to an organism and civilization to the organism's central nervous system. Humankind, in this image, would serve in the role of a geo-physiological administrator, tasked with managing the Earth System. Knowing how to manage an organization is to recognize what it needs, to register tiniest changes, and to react to them. I concur with Foss regarding the absence of genuine preference and immanent value from nature. Moreover, the neurological simile is certainly interesting. When the nervous system becomes dysfunctional, the organism suffers damage. Likewise, when civilization goes astray, climate change is the articulation of the damage done. But from a traditional Chinese perspective, the central nervous system is not the right metaphor for understanding the place and the function of human beings. This is still a Western way of thinking about nature. In the Chinese way, humans are centrally related to nature not because they are, or ought to be, planetary administrators managing the Earth system, but because, as mentioned, they have the ability to sustain the balance between Heaven and Earth. From a Confucian point of view, when human beings try to manage the natural world, they should not only apply their knowledge but also use their hearts. It is human beings with their virtuous behavior who are in the position to help nature in its transformation and sustenance. The neurological simile, by contrast, evokes a function or task of assuming central control over nature, which, arguably, is precisely what did cause the climate crisis in the first place.

Lao-Tzu provides a clear elucidation of the Tao, but it is doubtful whether his concept of non-action (*wu wei*, 無為) would assist us in dealing with the climate crisis. If Confucius' thinking can help us to change our attitudes and behavior, Confucian thought should be used as a mainstay of future environmental education.

In traditional Chinese culture, farmers and ordinary people always consult the lunar calendar whenever significant events happen. This custom has been practiced for thousands of years. For the past few years, though, farmers find that divination by the lunar calendar is no longer adequate for timing planting, plowing, growing crops and other agricultural activities. There is a proverb in China, that farmers' living depends on heaven. Now, it seems heaven cannot help them anymore. They lost their guidance.

## Conclusion

My discussion in this article covers three arguments I stated in introduction.

The ancient Chinese philosopher Confucius interprets the metaphysical concept of the Tao on the basis of the *I-Ching* in a way that discloses the ecological intelligence we need today. The Tao cannot be uttered, but all natural phenomena come from it. Confucius specifies that following the teaching of *I-Ching*, in addition to the two opposite poles, Qian and Kun, or Yang and Yin, six other trigrams that symbolize natural phenomena also emerge. This is one sense of 'I' (yì; 易) – it means simple and easy. Another operational principle is the principle of reversal. 'I' changes from one extreme to another and circles around. Many examples can be cited from the *I-Ching*. According to Confucius, if one has been cultivating virtuous sincerity, one can be part of the transformation and sustenance of the Tao, and the Human element can form a trinity with Heaven and Earth. This envisioned trinity is a stable and perfect form; it expresses a human-nature relationship that is sustainable and thus can last indefinitely.

The doctrine of Taoism is often regarded as the best example of ancient Chinese environmental thought. But a closer examination of the Taoist doctrine of wu-wei or non-action, as seen from a Confucian standpoint, makes one wonder whether it suffices for our present-day needs. Confucius, as a humanist, asks humankind to be involved actively in helping to transform and sustain Heaven and Earth. The teachings of the *I-Ching* guided his thoughts on how human beings should behave in the natural world.

We are familiar with many theories of environmental ethics. Most of them are developed by Western theorists and scholars, such as Leopold, Naess, Lovelock, Callicott, Wenz, Garvey and McCallum. In addition, there are also many theories with environmental significance that had been developed by scholars from the animal rights movement, ecofeminism etc. Comparing classical Chinese thought to these theories shows that Confucius already articulated insights of ecological intelligence, which civilization must apply in the age of climate change. Thus Confucius should be ranked as one of the pioneers of environmental thought.

*Text edited and revised by Martin Schönfeld*

## Notes

1. The commentary of *I-Ching* by Confucius, combined with his other commentaries of other ancient text of classics are called Ten Wings (十翼).
2. Confucius 1993, chaps 6 and 7 (易經繫辭傳上傳第六、七章).
3. Ibid., chap. 1.
4. The text of *I-Ching*, chap. 1.
5. Confucius, 2010a, chap. 15, #29 (衛靈公篇).
6. Ibid., chap. 17, #17 (論語，陽貨篇).
7. *Tao-Te-Ching*, 1994, chap. 5. I use the translation of M.H. Kwok et al.
8. Confucius 2010b, chap. 20.
9. Ibid., chap. 22 (中庸，第 22 章).
10. Some people translate this phrase as the sage or saints. For Confucius, this is honorific name for those who create the text of *I-Ching*. I would prefer using 'the greatest intellects'.
11. Xunzi said, 'Fire and water possess spirit but no life, grass and trees possess a life but no awareness, birds and animals possess awareness but no sense of morality, only humans possess spirit, life, awareness, as well as the sense of morality, hence the noblest beings in the world' (荀子王制篇第九).
12. Confucius 2010a, chap. 7, #1 (述而篇).

## References

Ames, R.T. 1989. Putting the Te back into Taoism. In *Nature in Asian traditions of thought*, ed. J.B. Callicott and R.T. Ames, 279–89. New York: State University of New York.

Capra, F. 1991. *The Tao of physics*. Boston: Shambhala Publications.

Cheng, C. 1998. The trinity of cosmology, ecology, and ethics in the Confucian personhood. In *Confucianism and ecology*, ed. M.E. Tucker and J. Berthrong, 211–35. Cambridge, MA: Distributed by Harvard University Press for the Study of World Religions.

Confucius. 1993. The commentary of *I-Ching*. In *The I-Ching*, trans. and ed. Y. Wu. Taipei, Taiwan: San Min Book Co., Ltd.

Confucius. 2010a. *Analects*. In *The four books*, trans. Chu-Si. Taipei, Taiwan: Chung Hwa Book Co.

Confucius. 2010b. *The doctrine of the Mean*. In *The four books*, trans. Chu-Si. Taipei, Taiwan: Chung Hwa Book Co.

Foss, J.E. 2009. *Beyond environmentalism: A philosophy of nature*. Hoboken, NJ: John Wiley.

Garvey, J. 2008. *The ethics of climate change: Right and wrong in a warming world*. New York: Continuum International Publishing Group.

Houghton, J. 2004. *Global warming: The complete briefing*. New York: Cambridge University Press.

*I-Ching*. 1993. Trans. and noted by Y. Wu. Taipei, Taiwan: San Min Book Co., Ltd.

Lovelock, J. 1995. *Gaia: A new look at life on Earth*. New York: Oxford University Press.

Lovelock, J. 2009. *The vanishing face of Gaia*. New York: Basic Books: A Member of the Perseus Books Group.
McCallum, I. 2008. *Ecological intelligence: Rediscovery ourselves in nature*. Golden, CO: Fulcum Publishing.
McKibben, B. 2007. *Deep economy*. New York: Henry Holt & Company.
Orr, D. 1992. *Ecological literacy: Education and the transition to a postmodern world*. New York: State University of New York Press.
Palmer, J.A. 1998. *Environmental education in 21st century: Theory, practice, progress and promise*. London: Routledge Press.
*Tao-Te-Ching*. 1994. Trans. M.-H. Kwok, M. Palmer, and J. Ramsay. Shaftesbury, Dorset, UK: Element Books Limited.
Tucker, M.E., and J. Berthrong, eds. 1998. *Confucianism and ecology*. Cambridge, MA: Distributed by Harvard University Press for the Study of World Religions.
Wenz, P.S. 2001. *Environmental ethics today*. New York: Oxford University Press.
Xunzi. 2010. Trans. and noted by Xiong Gong-Zhe. Taipei, Taiwan: The Commercial Press, Ltd.

# A Daoist response to climate change

Chen Xia[a] and Martin Schönfeld[b]

[a]*Chinese Academy of Social Sciences, Beijing, China;* [b]*Philosophy, University of South Florida, Tampa, FL, USA*

Climate change is now a global problem that can no longer be ignored. As climate change signals a civilization failure, the emerging reality will spur cultures everywhere to re-examine their traditions and rediscover the ecological wisdom of the ancients. Daoism will be no exception. This paper tries to explain the Daoist response to climate change by focusing on the manifestation of Dao, the responsibility of humankind and the ideal life. It shows that in its tenets and practices, Daoism articulates ideas that emphasize conservation, envision a post-consumerist existence and inform a climate ethics, which can help with the cultural adaptations that are now necessary for a sustainable future.

Li Gang (1998) notes that Daoism is a conceptual resource for solving environmental problems. And today, the ecological interpretation of Daoism has become a growing research area.[1] Thus, for example, Yin Zhihua (1996) examines the numerous measures to protect animals, plants, land, and water sources that are detailed in Daoist precepts. One of the authors of this article (Chen Xia) investigates the Daoist contribution to sustainability and ecofeminism (Yong and Xia 2001; Xia 2000). Such research is also done in the West, as the work of Tucker (1993), Clark (1984, 188), Devall and Sessions (1984, 100), Sylvan and Bennett (1988), Hall (1989), Ames (1989), and one of the authors (Martin Schönfeld, 2007, 2011) illustrates. For Tucker, Daoism contributes to a natural ecology. For Clark, the Daoist notion of 慈 (*ci*; 'loving kindness') has ecosophical potential. For Devall and Sessions, it is clear that 'sympathy, respect, and affection for life are the foundation of the Daoist life'. For Sylvan and Bennett, Daoism promotes the 'green life'. For Hall, Daoism assists in generating a new order of consciousness. For Ames, Daoism helps to solve key ecological problems, and for Schönfeld, Daoism informs the civil evolution that climate change demands.

This essay is an inquiry in Daoism and ecology in general and climate change in particular. In the investigation of Daoism and climate, we will attempt to re-interpret the relationship of the Dao to *wu* (物, 'things','matter','creatures').[2] Things and myriad things (萬 物, *wan wu*) are frequently mentioned in Laozi's *Dao De Jing* and in Zhuangzi's writings, the two chief texts of Daoism, but they are rarely discussed by scholars. What has not yet sufficiently been examined is the relation of *wu* to the standard array of Daoist terms, such as *dao* (道, 'the way'), *de* (德, 'virtue', 'power', 'potency'), *yi* (一, 'one'), *ziran* (自然, 'nature'), or *wuwei* (無為, 'non-action'). We suspect that such a contextual study of *wu* is the interpretive path for getting to a Daoist climate ethics.

## 1. Daoism and climate change

In recent decades, we have been witnessing record high temperatures, heat waves, erratic winters, violent storms, melting ice caps, thawing permafrost soils, and unbearable droughts in all parts of the world. The Intergovernmental Panel on Climate Change (IPCC), which is composed of researchers from over 130 countries, more than 2500 scientific expert reviewers, more than 800 contributing authors, and more than 450 lead authors, states in its Fourth Assessment Report (AR 4) 2007 that the warming of the climate system is 'unequivocal' (IPCC 2007, 2) The IPCC also states that 'since the start of the industrial era (about 1750), the overall effect of human activities ... has been a warming influence', and that 'the human impact on climate during this era greatly exceeds that due to known changes in natural processes'.[3]

The lion share of all greenhouse gases (GHG) was discharged in the collective pursuit of 'the good life'. Climate change is now a global problem that can no longer be ignored. Unless it is solved, we will not be able to ensure the future of humankind. We might be swept down a flooded road of incalculable losses. Because global warming is brought about by economic growth, consumption patterns, and lifestyles, we can say that this crisis is a human artifact. Without us climate would not change the way it does. We can further say that the worst of the changes can be headed off, at least for the time being. Some processes, such as the transformation of the Arctic, are irreversible, but elsewhere a window of opportunity remains open. The size of the window is continually shrinking, but it will never be completely shut. Even after reaching a systemic tipping point, we can still resolve to react, adapt, and survive.

The analysis of climate change is clear-cut. Civilization is the cause. The perpetration happens on three levels. A technological dimension manifests itself in the use of coal, oil, and gas as the GHG-producing fuels that power civilization. An economic dimension amounts to an ever-expanding market with ever more energy needs. A political dimension expresses itself in opportunistic servility to economic interests and in a willful disconnect from peer-reviewed data. Technology, economics, and politics are the joint cause of climate change.

Nature absorbs the effects as consequences that play out on three levels as well. A physical dimension reveals itself in the transformation of the Earth system, from a fertile planet to a world that will be hot, dry, and stormy. A geographical dimension discloses itself in changing coastlines, vanishing glaciers, and expanding deserts. A biological dimension manifests itself in the shifts of biomes, collapse of ecosystems, and overall reduction of biodiversity.

Civilization depends on nature. The triple consequences that unfold in the natural world result in a decline of biospherical services. Civilization is tied to the integrity of these services. Thus, nature relays the consequences of human conduct back to civilization. The cause of climate change accordingly involves a boomerang effect. When the boomerang strikes home, the blow felt by humans will be exposure and scarcity.

Empirical trends already document how this blow strikes home at present. Diminishing food supplies are first, falling water tables are next, and uninhabitable regions are just around the corner. The power to avoid the worst hinges on the three stated dimensions of civilization. Just as technology, economics, and politics currently feed into the collective perpetration of the crisis, they can eventually and concurrently contribute to its mitigation. Technological solutions such as scalable postcarbon energy sources and geoengineering present themselves. Economic solutions such as designing a sustainable post-growth market may be unavoidable. Political solutions such as creating and planning this new economy may have to be taken.

The etiology of the crisis spans a human perpetration of natural consequences that reverberate back to civilization, unless they are avoided by collective human solutions. Such an analysis, however, is comprehensive only on the surface. It omits the foundation. Next to the physical, geographical, and biological dimensions of the consequences, and in addition to the technological, economic, and political layers of perpetration and mitigation, there is another element, which is at the very heart at the issue. This is the cultural dimension. It consists of values and views, which are evident in conduct and expressed in choices. The cause of the crisis is grounded in choices – for instance, about whether to adopt one technological model, one economic system, and one political agenda over another. This cultural dimension is the basis of the crisis. And as M. Cano (2008) pointed out, since climate change is ultimately a cultural problem, its solution must be cultural, too.

Daoism offers such a cultural solution. As an indigenous religion with nearly two thousand years of history, it has shaped traditional Chinese culture. In its tenets, and practices, Daoism articulates ideas that emphasize conservation, envision a post-consumerist existence, and inform climate ethics. In contradiction to other spiritual views around the world, Daoism also harbors a special connection to climate as such. In the pre-industrial past, Daoist practitioners and sages paid attention to climate, because its shifts augured natural and social transformations. Daoist priests were often trained as stargazers and kept watch of celestial events and meteorological phenomena. Daoist teachings, about environment, existence, and society, can help with the cultural adaptions that are now necessary for a sustainable future.

As climate change signals a civilization failure, the emerging reality will spur cultures everywhere to re-examine their traditions and rediscover the ecological wisdom of the ancients. Daoism will be no exception. But it also enjoys an advantage over such traditions elsewhere. Compared with other forms of native spirituality and local knowledge on the globe, Daoism weathered the onslaught of missionary zeal and ideological fervor rather well. Colonialism failed to control its native areas, and this allowed it to avoid the onslaught of monotheism and its imperialist enforcers. Communism tried to suppress it during the Cultural Revolution but failed to destroy it as well, not the least because the tightly woven Daoist network in Taiwan remained out of reach. Other repositories of ancient ecological wisdom shattered in the clash with modernity and survived as mere fragments, but Daoism has retained its integrity.

But in this century, it is quite possible that Daoism will change. The cultural adaptations that prompt a quest for alternatives and a return to ancient wisdom will likely refine Daoism in turn. Environmental degradations have already triggered a surge of intellectual productivity from Daoist quarters, and it stands to reason that climate change will only boost this further. As T. Yi (2008, 132) put it, 'authentic wisdom never becomes outdated, but it has to be rediscovered and revived by each new generation on the strength of their experience and in their own usage'.

Daoism contains numerous precepts related to environmental conservation. Among them is the demand to care and even love plants, insects, animals, as well as mountains and rivers (Schipper 2001). Other stipulations concern the timing and the way of taking resources from the natural world. For example, during spring, when birds and beasts reproduce, raise their young, and are most vulnerable, hunting is prohibited, with sanctions against harming animals that are pregnant. With regard to the kingdom of plants, Daoist precepts only allow the harvesting of fruits and other vegetal sources that have reached maturity (Ming 1960, 572). There is a traditional sensitivity in Daoism about the protection of endangered species, which comes out in textual discussions of the number of species on the planet and the numbers of individuals left in a species. The Daoist concept of affluence, finally, refers to the largest number of species and individuals that an ecosystem naturally supports – affluence, in Daoist terms, is the state of a climax community.

Now, because the entire world is warming, rain forests, coral reefs, and other biomes are stressed and threatened. Daoism is concerned with species extinction, as the classic *Taiping Jing* makes clear: '(When) one thing is not brought to life, one lineage is discontinued; (when) many (things are not brought to life), then many lineages (are discontinued); (when) a few (things are not brought to life), then a few lineages (are discontinued)', and 'If one thing is incomplete or insufficient, then there is incompleteness in the heavenly way'.[4]

Daoist temples and organizations, meanwhile, promote the ethics of environmental protection. In 1995, the Chinese Daoist Association published *The Declaration of Chinese Daoist*

*Religion on Ecology and Environmental Protection*. More recently, nuns and monks in Daoist temples are devoting themselves to reducing their carbon footprint, by curtailing the use of incense and firecrackers, while expanding their energy independence, by installing solar roof panels and implementing zero-emissions design and 'green building' methods in the construction of new temples. In 2008, Daoist leaders met in Nanjing to pass a climate resolution, a seven-year plan of realizing sustainable designs and reducing carbon emissions in China.

## 2. The manifestation of Dao

Dao is the source of things and the guiding principle of the world. The Dao is not a thing. Laozi describes the Dao as the 'ancestor of myriad of things', the 'root of heaven and earth', the 'mother of the world, and as the "refuge for the myriad creatures"'.[5] One could describe Dao as the creative, evolutionary, and emergent ground of being. Upon creation Dao permeates its creatures. Laozi (verse 34) remarks that 'the myriad things depend on it for life yet it claims no authority'. Zhuangzi says: 'the Way is that whence the myriad things derive' (Mair 1994, 323).[6] The philosopher Hanfeizi (*ca*. 281–233 BCE) writes:

> The Dao is that by which all things become what they are. It is that with which all principles are commensurable. Principles are patterns (紋 *wen*) according to which all things come into being, and Tao is the cause of their being. Therefore it is said that Tao puts things in order (理 *li*). Everything has its own principle different from that of others, and Tao is commensurate with all of them [as one]. (Chan 1963, 260)

The classical collection known as the *Huainanzi* (2nd C BCE) contains another explanation:

> Mountains are high because of it. Abysses are deep because of it. Beasts can run because of it. Birds can fly because of it. The sun and moon are bright because of it. The stars and timekeepers move because of it. (Huainanzi 2010, 271; Ming 1960, 49)

As for *wu* (the things), Laozi (v. 29) says: 'some things lead and some follow; some breathe gently and some breathe hard; some are strong and some are weak; some destroy and some are destroyed'. Zhuangzi (c. 17, 42) says, 'There is no end to the weighing of things, no stop to time, no constancy to the division of lots, no fixed rule to beginning and end'. He adds (c. 19, 48), 'All that have faces, forms, voices, colors these are all mere things'. And he contrasts *wu* with Dao (c. 17, 43): 'The Way is without beginning or end, but things have their life'.

Dao and *wu* are opposites. *Wu* is transient, Dao is eternal; *wu* is structured, Dao is shapeless; *wu* involves distinct objects, Dao permeates all; *wu* is many, and Dao is one. Dao and *wu* are certainly different, but they are also inseparable. One can think of these opposites as poles along an evolutionary and ontogenetic continuum. They are joined in time, as a ground and its offspring, or as a creative power and its generated creatures. One can also think of these opposites as poles along an ontological continuum. In this perspective, they are joined in space, as simultaneous aspects of a unified reality. In that sense, *wu* is like matter, Dao is like energy, and Dao is *wu* just like matter is energetic. One can think of this co-presence analogous to a wet cloth – Dao surges through *wu* (道沖) just as water drenches a fabric.[7]

Zhuangzi (c. 22, 218) says: 'That which makes things [i.e. the Dao] has no boundaries with things, but for things to have boundaries is what we mean by saying 'the boundaries between things'. This illustrates the wet-cloth simile, for the boundless Dao penetrates all concrete things. Laozi adds:

> As a thing the way is,
> Shadowy, indistinct,
> Indistinct and shadowy,
> Yet within it is an image,
> Shadowy and indistinct,

Yet within it is a substance.
Dim and dark,
Yet within it is an essence.
This essence is quite genuine
And within it is a message that can be tested.[8]

Dim and vague as it is, Dao is not separated from the appearances. The Daoist view of reality is holistic, not dualistic as in Platonism or the Judeo-Christian belief systems. Daoism does not stress the separation of essence from appearances; on the contrary, it emphasizes the interrelatedness of everything. This has practical ramifications, as Daoism reminds people of their affinities and relations. Dao unfolds *wu* and thus precedes *wu*, but does not dwell on a transcendent plane or supernatural far side; Dao exist *through* the myriad things. When asked about the whereabouts of the Way, Zhuangzi (c. 22, 217) answers, 'there's no place where it is not present; even ants, the panic grass, tiles and shards have the Dao'. The philosopher Cheng Xuanying (seventh century CE) points out, 'Dao is not detached from *wu*, and *wu* is not detached from Dao. There is no *wu* beyond Dao; there is no Dao beyond *wu* (Xuanying 1988, 407)'

The ontological bond of Dao and *wu* means that the force of Dao dwells in the things. This transfer of power from the generative principle to the individuated creatures, which is unlike anything in the monotheisms in the West, implies that reality is dynamic, that nature is self-organizing, and that the *wu* are innately active. Laozi (v. 37) says, 'Dao never acts, but nothing left undone: if Lords and principle can hold fast to it, the myriad things will transform themselves'. The idea of *autopoiesis* or self-organization also implies that nature will always find a way if left to its own devices; exerting control is really not necessary. Zhuangzi (c. 11, 99) advises, 'merely situate yourself in non-action (無為, *wuwei*), and things will evolve of themselves', and adds ( Zhuangzi c.11, 99), 'do not ask its name; do not spy out its characteristics, things will assuredly come to life by themselves'.

Several things follow from this. First, as *wu* or the things of nature absorb Dao's power, dualistic distinctions between the secular and the sacred disappear. Similar to a Western deity, Dao deserves reverence and is essentially valuable. Unlike a Western deity, Dao is natural, not supernatural, and not a personified substance, but an energy flow. The flow manifests itself as a pulse, a beat, or a rhythm. The secular and the sacred merge, and nature gains intrinsic value in all its things. Second, as things absorb the Dao's power, and as this power is regarded as natural and good, nature's way is already the right way, and nature's power always finds its proper outlet. Human intervention is not needed. Leaving *wu* alone gives them the freedom to unfold. And finally, as *wu* express Dao, and as Dao deserves respect, protecting *wu* is our species' duty, as the only type of human intervention Daoism requests.

## 3. The responsibility of humankind

Climate change destabilizes the Earth system and ravages the biosphere. Without human emissions, these destabilizations and destructions would not have happened. The Earth system would have continued to pulse through its cycles, and the biosphere would have kept adding more biotic complexity from cycle to cycle. Anthropogenic influence on the natural flow of things is thus jarring, destructive, and goes against nature's flow.

The proper human role is to aid and abet nature's flow, not to go against it. There is also a practical reason for this, for only when humans act in accordance with the Dao, they can enjoy living together with other *wu* in a peaceful, healthy, and prosperous way. This role is also illustrated in the Daoist proverb that humanity is 'the universe's elder brother who looks after the myriad things' (*li wanwu zhi zhang* 理万物之长). The rise and fall of things partly depend on humans. For their creation, things depend on the cosmos, for their root, they depend on the Earth, and for their security, they depend on humans (Ming 1960, 174). Laozi takes it for granted that humans ought to be involved in Dao's self-generating process, and this is why he says,

> Hence the way is great;
> Heaven is great;
> Earth is great;
> And human is great.[9]

Daoism assumes that all living creatures and humans maintain a kind of balance, depend on each other, and make use of their shared concerns in the space between heaven and earth. This factual assumption translates to a value prescription: just as this is how it already is, we ought to abide by it – we should maintain balance, we should depend on one another, and we should make use of our shared concerns. The reason for this imperative is a practical worry. When humans upset this balance, cut the connections, and split from shared concerns, they make themselves, as well as heaven and earth, prone to sickness. There will be pestilential airs (*qi* 气) in the world.

The *Taiping jing* describes this particular outlook as follows:

> The heavenly signs (i.e., astronomical phenomena) and earthly patterns are in disorder, and thus heaven and earth become ill. This makes the three luminaries (i.e., the sun, moon, and stars), wind and rain, the four seasons and the five agents fight. It will be an inauspicious year for crop harvest. (Ming 1960, 356)

The unprecedented warming of the global climate system has been variously described as a 'fever' that we are giving the planet. This fever is symptomatic of an observable and quantifiable illness, whose progression amounts to the destruction of the natural supports of our civilization, with food becoming the first weak link.

Helpful to understanding the proper human role is the Daoist dyad of the uncarved block (*pu* 朴) and the vessels (*qi* 器). Laozi (v. 28) says, 'when the uncarved block shatters it becomes vessels'. One could read the uncarved block as the Dao and the vessels as the *wu*. The basic precept of Daoist ethics is that humans ought to act in accordance with the Way. Just as the Dao turning into *wu* and the uncarved block shattering into vessels, humans should process and transform things for their use. But such an activity should not violate their original nature. What humans should do is to disclose, to bring out, to actualize potentials, and to bring forth.

Humans have the right to use, process, and transform things, and the duty to follow the natural flow of events. Activities should not disrupt natural processes, not upset biospherical balances, and generally remain types of non-action. The Daoist precept of non-action or *wuwei* (無為) does not mean to do nothing. It can be translated as 'effortless action' or 'non-calculating action'. In Laozi it often appears together with the notion of 'nothing left undone' or *wubuwei* (無不為). Non-action could be described as acting in a manner that is as gentle and non-violent as the Dao, and that is the opposite of acting in a compulsive, reckless, and willful way. In Daoism, *wuwei* is the formal precept of practice. A.C. Graham (1989, 232, 233) writes that it is essential 'not to interfere when things are already running well by themselves.... [and to] maintain the concord by minimal interference to maximum effect, discerning the incipient danger before it develops and 'doing to it before it is something'.

For climate ethics, Daoism accordingly specifies two responsibilities for humankind, one positive and another negative. In positive terms, Dao brings forth life and complexity, and humans should do the same. Their first responsibility in a changing climate is to learn to become stewards of the biosphere, to shepherd life, and protect complexity. In negative terms, Dao acts by non-acting; that is, by acting in harmony with the natural flow. Their second responsibility is accordingly to refrain from further disruptions of the flow, and to learn to become mitigators of climate change, to soften the impact and to calm down the waves.

## 4. The ideal life

The main difficulty that prevents humankind from assuming its responsibility is the unsustainable pattern of production and consumption, especially in the highly developed countries and among the wealthy oligarchs in the developing world. The data speak for themselves. Today, as little as 26% of the world population, in the industrialized nations, consume 75% of the world's energy and 80% of the world's resources, while releasing 80% of the world's GHG emissions. The remaining 74%, spread over more than 100 countries, consume only a quarter of the world's energy, a fifth of the world's resources, and are responsible for a mere 25% of GHG emissions. This contrast between the polluting rich and the frugal poor is not the only division in humankind. Perhaps even more troubling is that the rich themselves split into those who at least try to aspire to some sort of sustainable future and those who roundly reject any sorts of visions. Here the statistics are as well known as they are notorious: the USAs amounts to less than 5% of the global population, but is responsible for more than a third of the world's GHG emissions. US citizens, on an average, enjoy a comparable standard of living as those of the European Union while using twice the amount of energy per capita. Even the specifics are astounding: the USA 'consumes more gasoline than the next 20 countries combined (including Japan, China, Russia, Germany, and Brazil). The USA – with 249 million passenger vehicles out of the global 912 million – not only has the largest fleet but it is near the top in miles driven per car and near the bottom in fuel efficiency' (Brown 2010).

This unsustainable pattern, interestingly, also fails to reach its self-asserted goal. There is now useful and rigorous empirical research on happiness. It is recognized that above subsistence levels, wealth has no direct relation with happiness.[10] The consumer culture of the USA, the most excessive and wasteful in the world, is also a society that ranks lower in happiness than others do (Layard 2006; McKibben (2007, 41–2). Obviously, then: for the interest of our descendants, the survival of contemporary poor people, the preservation of endangered species, the stabilization of the climate, and, last but not least, for our own rational self-interest, it would be best to change our consumption and emission patterns and to choose voluntarily a simple lifestyle.

Laozi (v. 19) says: 'not to value goods which are hard to come by will keep them from theft'. He also says (v. 67), 'I have three treasures which I hold and cherish. The first is known as compassion, the second is frugality, the third is known as not daring to take the lead in the empire' (Laozi, v. 67). The Daoist adept or the 'perfect man' (*zhenren* 真人) often described by Zhuangzi seeks to wander in the world as a free and independent spirit, and lives a life of poetic sentiment detached from the material possessions. The ideal life, for Daoists, is the simple life.

The Daoist goal of the simple life is reinforced by the contemporary context of consumerism. To promote this goal today is different from promoting it in the past, just as it is different in the highly developed world as it is in the developing nations. It is one thing to live frugally when you have to, but it is another if we can choose to do so. The value of a voluntary return to a life of simplicity and authenticity is all the higher because it appears its adoption appears as the only rational response to the unsustainable pattern that has unleashed the climate crisis. In this sense this return also appears as the next step of human development, making consumerism a transient stage, and revealing simplicity as the ultimate goal of civil evolution.

A further reason why the return to simplicity is the ideal life concerns the reconstruction of human links to the larger whole. Simplicity affords the individual an existence that is more in tune with nature. For Daoism, the universe is an enormous body full of vitality. The *Huainanzi* states, 'Heaven and Earth, space and time, are as the body of a single person' (Huainanzi 2010, 271). Indeed, in Daoist medicine, the human body is conceived as having a connection with the greater cosmos. Humans and nature not only are connected by mutual exchange and influence, but also belong to the same organic whole. This connection is not a mere composition, or an integration of a part into an aggregate, but also a mirror: humans constitute a microcosm and nature a macrocosm. The Daoist image of the Earth as the corporeality of nature is articulated in this way: 'Spring are the blood of the earth; rocks are the bones of the earth; fertile soil is the flesh of the earth' (Ming 1960, 120).

So, in a wider sense, humans and nature share the same body. This allows humans to free themselves from the constraints of individuality and meditatively fuse with the greater body of the universe. Honing the relationship between human vitality and the pulse of the cosmos is a Daoist contemplative practice. This may suggest that simplicity, as the ideal life, is a stepping stone to something even greater. But be this as it may, climate change is a visceral reminder that *we are in the world*; that our being is part of a larger being, and that going against the flow of nature means to run counter our collective self-interest. Daoism provides an alternative model of being-in-the-world, and in contrast to the unsustainable practices that shape consumerist civilization, this alternative is universalizable, in a Kantian sense: if everyone embraced this alternative as a maxim, turning it into new pattern of civilization, this pattern would not be self-reducing – it would not collapse under its own weight but instead simply allow us to continue, evolve, and flourish as a civilization.

## Acknowledgements

The authors thank Kimberly Powers and Grégoire Espesset (École Pratique des Hautes Études, Institut des Hautes Études Chinoises, Paris), for help with translations, especially from the Taiping jing.

## Notes

1. See, for instance, Zhang Jiyu 张继禹, Li Yuanguo 李远国, and Zhang Xingfa 张兴发, Daofa ziran yu huanjing baohu 道法自然与环境保护 [Daoist natural law and environmental protection] (Beijing, 1998); Le Aiguo 乐爱国, Daojiao shengtaixue 道教生态学 [Daoism and Ecology] (Beijing, 2005); Mao Liya 毛丽娅, Daojiao yu Jidujiao shengtai sixiang bijiao yanjiu 道教与基督教生态思想比较研究 [Daoism and Christianity: Comparative research on ecological thought] (Chengdu, 2007); Cai Linbo 蔡林波, Zhu tian sheng wu: Daojiao shengtaiguan yu xiandai wenming 助天生物: 道教生态观与现代文明 [Helping Heaven's creatures: Daoist ecological perspectives and modern civilization] (Shanghai, 2007); Chen Xia, Chen Yun, and Chen Jie, eds., 道教生态思想研究 [Research on Daoist ecological thought] (Chengdu, 2010).
2. There are two standard spellings of the Chinese term '道', commonly translated as the Way: 'Dao' in the modern pinyin Romanization, and 'Tao' in the older Wade-Giles system. We will use pinyin throughout, but retain the older spelling in citations if needed.
3. IPCC Working Group 1 (2007, 5): 'Most of the observed increase in global average temperatures since the mid-20th century is very likely due to the observed increase in anthropogenic GHG concentrations'.
4. Wang Ming (1960, 219) (from the Taiping jing chao, juan 4, not the Taiping jing proper) and p. 462 (from the Taiping jing, juan 102, pian 166). Translation by Grégoire Espesset.
5. Laozi, verses 4, 25, and 62. All Laozi references cited in this paper are from Lao Tse, *Tao Te Ching*, trans. D.C. Lau (Hong Kong: Chinese University Press, 1982), unless noted otherwise.
6. All citations from Zhuangzi, by chapter and page number, refer to Mair's edition.
7. Compare Laozi, v. 4
8. Laozi, v. 21. I changed 'something' into 'message' here. D.C. Lau translates 其中有信 as 'within it is something that can be tested'. 信 (xin) means 'message', following Wang Bi's reading; cf. Lou Yulie, Collected works of *Wangbi* 王弼集校释 (Beijing: Zhonghua Publishing, 1980), 53
9. Laozi, v. 25. Here, the term for 'human' replaces the term for 'king'. The ancient Fu Yi and Fan Yangyuan's edition list 'human' here. In the received edition, after 'king counts as one' follows 'human models himself on Earth/Earth on heaven/Heaven on the way/And the way on that which is naturally so'. 'Human, Earth, Heaven, Way' reverses the one given in the previous paragraph 'Way is great/

Heaven is great/Earth is great/Human is great'. The concept 'king' does not work in this context; cf. Chen (2003, 171–2).

10. 'Money consistently buys happiness right up to about $10,000 per capita income, and after that point the correlation disappears' (McKibben 2007, 41; Diener and Seligman 2004, Figure 2, 5).

## References

Ames, Roger T. 1989. Putting the Te back into Taoism. In *Nature in Asian traditions of thought: Essays in environmental philosophy*, ed. J. Baird Callicott and Roger T. Ames, 113–44. Albany: SUNY.

Brown, Lester. 2010. *Plan B 4.0. Mobilizing to save civilization*. New York: Norton.

Cano, Marcel (University of Barcelona). 2008. Cultural dimensions of climate change symposium, meeting of the Society for Human Ecology, Western Washington University, Bellingham, WA, September.

Chan, Wing-Tsit, ed. 1963. *A source book in Chinese philosophy*. Princeton: Princeton University Press.

Chen, Guying. 2003. *Lao-zi jinzhu jinyi*. Rev. ed. Beijing: Commercial Press.

Clark, John. 1984. *The anarchist moment: Reflections on culture, nature and power*. Montreal: Black Rose.

Devall, Bill, and George Sessions, eds. 1984. *Deep ecology: Living as if nature mattered*. Salt Lake City, UT: Shambhala.

Diener, Ed, and Martin E.P. Seligman. 2004. Beyond money: Toward an economy of well-being. *Psychological Science in the Public Interest* 5: 1–31.

Gang, Li. 1998. The meaning of Daoism in modern world. *Journal of Religious Studies* 1 [Chinese]: 1.

Graham, A.C. 1989. *Studies in Chinese philosophy and philosophical literature*. Albany: SUNY Press.

Hall, David L. 1989. On seeking a change of environment. In *Nature in Asian traditions of thought: Essays in environmental philosophy*, ed. J. Baird Callicott and Roger T. Ames, 99–111. Albany: SUNY.

Huainanzi. 2010. *The Huainanzi: A guide to the theory and practice of government in Early Han China*. Trans. J.S. Major et al. New York: Columbia University Press.

IPCC (various authors). 2007. *Climate change 2007: Synthesis report – summary for policy makers*. http://www.ipcc.ch/pdf/assessment-report/ar4/syr/ar4_syr_spm.pdf (accessed March 20, 2011).

IPCC Working Group 1 (various authors). 2007. *Climate change 2007: The physical science basis*. Frequently asked questions, p. 100. http://www.ipcc.ch/pdf/assessment-report/ar4/wg1/ar4-wg1-faqs.pdf (accessed March 20, 2011).

Mair, Victor H. 1994. *Wandering on the way: Early Taoist tales and parables of Chuang Tzu*. New York: Bantam.

McKibben, Bill. 2007. *Deep economy*. New York: St Martin's.

Ming, Wang. 1960. *Taiping jing hejiao*. Beijing: Zhonghua Publishing.

Schipper, Kristopher. 2001. Daoist ecology: The inner transformation. In *Daoism and ecology: Ways within a cosmic landscape*, ed. N.J. Giradot, James Miller, and Xiaogan Liu, 79–84. Cambridge, MA: Center for the Study of World Religions.

Schönfeld, Martin. 2007. How real is Daoism? Making sense of Laozi, Kant, and climate. *International Journal for Field Being* 6, 10 pp. Reprinted in Lik Kuen Tong, ed. 2009. *Wei Wu Wei: Essays on Taoist philosophy*. Hong Kong: Daoist Association Press, 2009.

Schönfeld, Martin. 2011. Daoism and climate change. *Daoism Studies*. Forthcoming.

Sylvan, Richard, and David Bennett. 1988. Taoism and deep ecology. *Ecologist* 18: 148–59.

Tucker, Mary Evelyn. 1993. Religious values derived from other traditions: The ecological dimensions of Taoism and Confucianism. *Dialogue and Alliance* 7: 95.

Xia, Chen. 2000. On the Daoist feminine and eco-feminism. *Journal of Southwest University of Nationalities* 8 [Chinese]: 37–40.

Xuanying, Cheng. 1988. *daode zhenjing xuande zuanshu* in *Daozang*. Vol. 13. Beijing: Wenwu Publishing.

Yi, Tang. 2008. *Breeze from a deep valley*. Hangzhou: Zhejiang University Press.

Yong, Chen, and Chen Xia. 2001. Daoism and sustainable development. *Journal of Religious Studies* 3 [Chinese]: 45–51.

Zhihua, Yin. 1996. Environmental conservation in Daoist precepts. *Daoism in China* 2 [Chinese]: 33–4.

# Justice, negative GHIs, and the consumption of farmed animal products

Jan Deckers

*Institute of Health and Society, The Medical School, Newcastle University, Newcastle-upon-Tyne, NE2 4AX, UK*

In a previous work, I argued that all human beings should possess the right to adequate health protection and that we have good reasons to believe that not all human beings are or will be able to enjoy this right. I introduced the 'Global Health Impact' or 'GHI' concept as a unit of measurement to evaluate the effects of human actions on the health of human and nonhuman organisms and argued that the negative GHIs produced by our current generation jeopardise the right to adequate health protection of future generations unjustifiably. In this article, this incomplete theory of human justice is applied to the negative GHIs associated with the consumption of farmed animal products. Since the negative GHIs of such products generally exceed those associated with other diets, I argue that those who exceed their fair share of negative GHIs might curtail their consumption and that the costs of reducing negative GHIs in this domain might be lower than the costs associated with other things that could be done to reduce negative GHIs.

## Introduction

Elsewhere, I introduced the 'Global Health Impact' (GHI) concept as a unit of measurement that evaluates the effects of human actions on human and nonhuman health (Deckers 2010b, 2011b). I also argued that the total negative GHIs of our present generation jeopardise unjustifiably the right to health protection that should be possessed by all human beings, including those who will belong to future generations. Consequently, I argued that our present generation is under a moral obligation to limit its negative GHIs and that this obligation must be borne primarily by those who exceed their fair share of negative GHIs. In this article, this incomplete theory of human justice is used to evaluate which duties might follow from an ethical evaluation of the human health costs and benefits associated with the consumption of farmed animal products.

Several studies have emphasised the role that animal products can play in the human quest for adequate nourishment (Eik et al. 2008; Herrero and Thornton 2010). Since animal products are relatively energy-dense, rich in high-value protein, vitamin B-12, and fat-soluble vitamins, absorbable minerals such as iron and zinc as well as other micronutrients, some have welcomed their inclusion in the diets of people who might otherwise lack adequate nutrition (Leroy and Frongillo 2007; Müller and Krawinkel 2005). However, in recent years, the consumption of farmed animal products has also gained increasing recognition as a human health concern. One concern is its connection to obesity. The fact that more than one billion people are now overweight has prompted some to speak of an 'obesity pandemic' (McMichael et al. 2007). A recent study by Lloyd-Williams et al. (2008) estimates that even small reductions in the intake of animal products within many countries of the European Union could have significant positive health advantages. The researchers found that reducing saturated fat consumption by 1% and increasing monounsaturated and polyunsaturated vegetable fat consumption by 0.5% each would result in 'approximately 9800 fewer coronary heart disease deaths and 3000 fewer stroke deaths each year' in 15 countries of the European Union (Lloyd-Williams et al. 2008, 535). In general, diets that include small quantities of, or no animal products have been associated with lower levels of cardiovascular diseases, obesity, dyslipidaemia, hypertension, type 2 diabetes, and some types of cancer (Messina and Burke 1997; World Cancer Research Fund and American Institute for Cancer Research 2007). Additionally, the claim has been made that – if carefully chosen – adopting such diets does not produce adverse health effects (WHO 2003). In spite of this, the consumption of farmed animal products remains very high in many Western countries, whereas consumption levels in many other countries, although still modest, are rising at unprecedented rates, with an increase in the consumption of meat from 73 million tonnes in 1990 to 137 million tonnes in 2002 in 'developing countries' (Steinfeld et al. 2006, 15).

Apart from containing saturated fats, many farmed animal products include many other substances that are harmful to human health, including antibiotics, hormones, pesticides, dioxins, and metal compounds. Most animals are given antibiotics not because they are ill, but because they live in close proximity to other animals who might be ill or because it makes them grow faster (Anomaly 2010). Hormones are also given widely to stimulate growth or milk production. Animal products also contain relatively high levels of some pesticides because some of these concentrate up the food chain, a process known as bioaccumulation. The same applies to dioxins (such as polychlorinated dibenzodioxins and polychlorinated dibenzofurans) and polychlorinated biphenyls, which are causing a variety of health problems, including cancers and immunologic and metabolic disorders (National Academy of Sciences 2006; Swedenborg et al. 2009). Whereas dioxins are produced mainly by the incineration of medical and other wastes, including some plastics, and by some methods used to bleach wood pulp, they enter the human body largely by their ability to accumulate in the fatty tissues of animals that are then consumed by humans. Animals can be exposed to these harmful substances in a variety of ways, including their ability to travel in the air and to land on plant foods that are then eaten and through their presence in rendered animal fats that are used to feed animals (Sapkota et al. 2007, 666). Dioxins and methyl mercury can also be stored inside the bodies of fish and pose health risks to humans who eat them (Trasande, Landrigan, and Schechter 2005). The feed that pigs and birds who are consumed by humans eat frequently contains organoarsenicals such as roxarsone, which is used to increase feed efficiency and to stimulate growth (Chapman and Johnson 2002). Both inorganic arsenite ($As^{III}$) and inorganic arsenate ($As^{V}$) – human carcinogens – can thus be produced inside the animals' digestive tracts and their waste (Sapkota et al. 2007, 665).

Common bacterial pathogens include Salmonella, Listeria, Toxoplasma, *Escherichia coli* 0157:H7, various species of Campylobacter, *Clostridium botulinum* (the pathogen that causes botulism), and Hydatid disease. Parasitic diseases include Giardia, Cryptosporidium, Microsporidia, Fasciola, tapeworms, and threadworms. Many bacteria (such as *E. coli* 0157:H7 and Salmonella) are now becoming resistant to the drugs that have been developed to fight them, and some antibiotic-resistant bacteria have been transmitted from pigs and birds to humans (Aarestrup et al. 2000; Burkholder et al. 2007; Duffy and Moriarty 2003). Many viral diseases also emerge within the farm animal industry, including various viral haemorrhagic fevers, and the Nipah virus. Another disease is the Creutzfeldt–Jakob disease, which can spread not only through eating parts of animals carrying bovine spongiform encephalopathy, but also through blood transfusions and the use of contaminated surgical and other equipments (Collinge 1999). Whereas diets that exclude farmed animal products are not entirely free from all these health hazards, the nutritionist Sanders (1999, 265) has claimed that 'most food-borne disease in developed countries is transmitted via meat, eggs, shellfish, and milk'.

Given the wide range of human health concerns that are associated with the consumption of farmed animal products, of which I have provided no more than a concise overview, it seems surprising that relatively little attention has been given to the questions of whether and when the consumption of farmed animal products might survive moral scrutiny in relation to the goal of safeguarding human health. A possible reason for this deficit may be the possibility that many people hold the erroneous view that people who consume harmful farmed animal products primarily harm themselves. The impact of this view on public policy would be particularly strong within liberal societies that give prominence to the view that self-harm should not be given the same level of moral scrutiny when compared with the harm that might be inflicted on others, especially when combined with a strong anthropocentric culture that disregards any harm that humans inflict on nonhuman organisms. However, it is becoming increasingly clear that the consumption of farmed animal products can have far-reaching implications for the health of other human beings. This is exemplified by the emergence and spread of zoonoses, the diseases that are transmitted from nonhuman to human animals, which pose profound ethical questions that have enjoyed very little debate (Deckers 2011a). A review of 1407 species of human pathogens found that more than half were zoonotic and the same review found that zoonoses account for almost three-quarters of recently emerged human diseases (Woolhouse and Gowtage-Sequeira 2005). Particularly noteworthy are the strains of swine and avian influenza that could result in significant negative health consequences, even for those who play little or no part in their emergence. Whereas a variant of the avian influenza virus may have caused the great influenza epidemic that emerged in 1918, there is real concern that the recently emerged H5N1 and H1N1 strains could affect millions of people living today (Belshe 2005; Fraser et al. 2009). To emphasise this point, Greger (2007, 277) has commented that the 'same "trench-warfare" conditions' that existed in 1918 'can be thought to exist at present in many respects in industrial poultry production'. The conditions in which many pigs are farmed might not be much different. In addition, the transmission of these viruses, as well as of many other harmful pathogens, may be enhanced by the increasing urbanisation and globalisation of our increasing human population.

Whereas these issues seem sufficiently important to demand greater scrutiny of the consumption of farmed animal products, those who have questioned current consumption and production patterns out of human health concerns have done so predominantly out of concern for the more indirect ways in which the consumption of farmed animal products might jeopardise human health, with a particular focus on climate change and our use of resources. Particularly noteworthy in this respect is an influential report with the title 'Livestock's long shadow', published in 2006 by the Livestock, Environment, and Development Initiative, a group coordinated by the Food and Agriculture Organization of the United Nations (henceforth: the LEAD study) (Steinfeld et al. 2006). The report claims that the 'livestock sector enters into more and direct competition for scarce land, water and other natural resources' and that 'the environmental impact of livestock production will worsen dramatically ... in the absence of major corrective features' (Steinfeld et al. 2006, xxi, 275). Whereas the authors of this report do not use their findings to argue for a reduction in consumption, but predict that it will rise to unprecedented heights, their findings have inspired others to do so (Compassion in World Farming 2007; Garnett 2009; Goodland and Anhang 2009). Recently, the author of an influential review on the economic costs of climate change, Stern (2006), is reported to have claimed that vegetarian diets should be advocated on the basis of the view that farm animals put great pressure on our resources and that dietary changes must be considered 'for the world to break free from its catastrophic current trajectory' (Pagnamenta 2009). If these claims are correct, we have good reasons to be concerned about the farm animal sector if we want to safeguard the right to adequate health protection for all human beings. In what follows, I shall examine these claims to address whether, by curtailing their consumption of farmed animal products, people might be able to reduce their negative GHIs where they are required to do so by the demands of justice.

### Farmed animal products, greenhouse gas emissions, and negative GHIs

Whereas the LEAD study claimed that the farm animal sector produced 18% in $CO_2$ equivalents of all anthropogenic greenhouse gas emissions in 2002 (Steinfeld et al. 2006), a more recent study claims that the data provided by the LEAD study were a gross underestimate and that the total emissions of the sector amounted to 51% in $CO_2$ equivalents in 2009 (Goodland and Anhang 2009). The main reasons for this significant difference are attributed to the following: that the former study did not include respiration as a source of emissions; that it did not factor in the opportunity costs associated with the fact that a lot of land (26% of grassland and 33% of arable land) that is used by the farm animal sector could regenerate as forest and capture much more carbon through photosynthesis; that it undercounted the number of farm animals (for example by excluding farmed fish) and overlooked some emissions produced by the production, distribution and disposal of animal products, their by-products, and their packaging; that it ignored the emissions produced by the pharmaceutical and medical industries in their fight against diseases associated with the farm animal sector; and that an inappropriate 'global warming potential' (a measure that calculates the amount of heat trapped by a quantity of gas as a factor of the heat trapped by one unit of a similar mass of $CO_2$ gas) of 23, rather than the more appropriate figure of 72, was used for methane. With regard to this last reason, the authors justify their figure by pointing out that a 20-year timeframe (with global warming potential of 72) must be used rather than a 100-year timeframe 'because of both the large effect that methane reductions can have within 20 years and the serious climate disruption expected within 20 years if no significant reduction of greenhouse gases is achieved' (Goodland and Anhang 2009, 13). I have reported elsewhere how emissions from the farm animal sector will escalate in the future if current emissions data from the latter study are combined with the growth projections of the former study (Deckers 2010a). While I shall not discuss which of these studies is most likely to be correct, it is clear that both studies agree that the farm animal sector presents significant human health concerns because of its contribution to climate change.

To assess how dietary changes could reduce overall emissions, however, we need to develop some idea of how the emissions produced by different diets compare with one another. In this respect, a study carried out in the USA revealed that the mean diet of a US citizen, which includes 27.7% of calories from animal sources (comprising 41% from dairy, 5% from eggs, and 54% from a range of meats), produces at least 1.5 t more emissions (measured in $CO_2$ equivalents) per year, compared with the emissions produced by a vegan USA citizen (Eshel and Martin 2006, 13). To obtain some idea of how this compares with the emissions produced by personal transportation, the authors point out that the per-capita car miles travelled by the average US citizen in 2003 were 8332 miles, producing between 1.19 and 4.76 t of $CO_2$ emissions, depending on which vehicle was used (Eshel and Martin 2006, 2–3). The authors draw on their knowledge of the emissions produced by different car models to make an interesting analogy. If we imagine that a person adopting the mean diet of the US citizen drives an averagely efficient car (Toyota Camry) and that a vegan compatriot drives one of the most energy-efficient hybrid vehicles (Prius) on the US market in 2006, the difference in diet-related emissions (for a given quantity of food with equal caloric intake) amounts to the difference in emissions produced by the former driving 143 miles in the less-efficient car and the latter driving 100 miles in the more-efficient car (Eshel and Martin 2006, 2–3). Or, to use a different analogy, the difference in emissions between the person adopting the mean US diet and the person adopting the vegan diet also corresponds to the difference in emissions between driving 8332 miles (the distance travelled by the average US citizen in a year) in one of the most efficient cars and not driving at all.

If we now assume that an (not very) enlightened omnivore would be willing to use no more than half of the animal products consumed in the mean US diet and that the proper and large-scale use of new technologies could reduce the greenhouse gas emissions related to the production of farm animal products by 20%, enlightened omnivores living in the USA would still produce 0.6 t more greenhouse gas emissions per year when compared with their vegan compatriots (which comprise no more than 1.4% of the population according to a poll carried out in 2006) (Stahler 2006). Similar results were found in a more recent Swedish study, where the greenhouse gas emissions of three Swedish meal options were compared. The difference between the hypothetical vegan meal and the two hypothetical meals that included animal products varied between a factor of 3 and a factor of 11, depending on which kinds of animal products were chosen (Carlsson-Kanyama and González 2009, 1708S). In spite of these significant differences, the authors comment that 'research is needed to understand why dietary change is not on the climate change agenda' (Carlsson-Kanyama and González 2009, 1704S). I argue the more important moral point that diet should be on our moral agenda and that it is highly relevant when it comes to determining what our duties are in relation to climate change. To avoid dangerous climate change, which – so it has been argued – would violate human rights (Bell 2011; Caney 2009), the Intergovernmental Panel on Climate Change (2007, 20) has claimed that a reduction in $CO_2$ emissions of up to 85% relative to 2000 levels may be required by 2050. On this basis, Rosales (2008, 1414) has calculated that this leaves no more than an average of 0.8 t per capita per year, a threshold that many people exceed by their diet-related emissions alone.

It is on the basis of the greenhouse gas emissions produced by different diets that a research team led by McMichael has proposed that the global per-capita consumption of meat should drop from the current average of 100 g per person per day to an average of 90 g of meat per person per day (with less than 50 g coming from red meat from ruminants) (McMichael et al. 2007). Henceforth, I simply refer to this team by the name of McMichael. McMichael argues that all countries should eventually converge to a per-capita consumption level of 90 g per person per day, which would stabilise greenhouse gas emissions from the farm animal sector by 2050 relative to its contribution in 2005, subject to the global human population not having increased by more than 40% by 2050 and current emissions being reduced by 20% per unit of meat.

This proposal raises a number of concerns. One concern is the question whether a 20% reduction per unit of production could be achieved. This has been doubted as it would require the rapid, full implementation of all available technologies on a global scale (Weidema et al. 2009). A further issue is that the expected reduction would not occur unless it would be accompanied by similar reductions in the consumption of all other farmed animal products apart from meat. Another issue is that this proposal fails to take into consideration the different circumstances in which people living in different countries may find themselves. Some countries may rely more on the consumption of farmed animal products because some of their inhabitants may lack sufficient alternatives that are suitable for human consumption, which questions the desirability of the view that consumption levels between people living in different countries should eventually converge. A more fundamental problem is the question why McMichael only seeks a stabilisation of emissions from the farm animal sector, whereas many have argued that significant cuts in overall emissions are required to avoid dangerous climate change. For example, by passing the Climate Change Act 2008, the UK Parliament (2008) adopted the view that a reduction of at least 80% is required by 2050. Also, both the Climate

Change Act 2008 and the Kyoto Protocol take the 1990 emissions as the baseline level, whereas McMichael's team merely seeks for emissions to stabilise relative to the 2005 level. This reduces the contribution of this sector to climate change abatement strategies even further, especially since the global consumption of farmed animal products increased significantly between 1990 and 2005. Indeed, the LEAD study has estimated that whereas the consumption of meat levelled off in 'developed countries', total meat consumption in 'developing countries' almost doubled between 1990 and 2002 (Steinfeld et al. 2006, 15).

While it is, therefore, unlikely that McMichael's proposal goes far enough, it must be recognised that the fact that many people's diets are associated with relatively large amounts of greenhouse gas emissions need not necessarily be a moral problem. The issue of what counts as a fair greenhouse gas allowance should be considered as an important part of, yet not as a substitute for, the more general question of what counts as a fair threshold of negative GHIs. Likewise, the issue of what counts as a fair diet should not be treated in isolation from the question of what counts as a fair threshold of negative GHIs. Many people may well prefer either to reduce their negative GHIs elsewhere, for example, by reducing the emissions produced by the consumption of domestic energy or transport, or to offset some of their negative GHIs by producing positive GHIs, for example, by planting trees. Therefore, some people might justifiably be allowed to emit a relatively larger share of greenhouse gas emissions or to eat more farmed animal products than the amount proposed by McMichael, provided that their threshold allowance of negative GHIs is not exceeded.

## Farmed animal products, dietary ecological footprints, and negative GHIs

To obtain a better idea of the negative GHIs associated with the consumption of farmed animal products, it is clear that we must move beyond the question of how many diet-related greenhouse gas emissions each person produces. The ecological footprint concept might be useful here (Wackernagel and Rees 1996). Whereas the GHI concept measures the impact of human actions on human health in one unit, the 'ecological footprint' concept measures the impact of human activities on the nonhuman environment in one common unit, the use of bioproductive space.

Ecological footprint calculators have been developed to estimate the ecological footprints of different categories of individuals living in different parts of the world. The Global Footprint Network, an organisation set up in 2003 to promote sustainability, is updating its footprint calculator at the time of writing, and therefore data can only be obtained for people living in a few countries. If the calculator is used, for example, to calculate the footprint of a US citizen who lives in a free-standing home with running water and electricity where three people live together and where mean values (as indicated) are chosen for all remaining categories, a marked difference can be observed between the ecological footprint of a male person who eats 'meat a few times a week and eggs/dairy almost daily' and a male person who never eats animal products. The former would use about 8.5 global ha (21.1 global acres), whereas the latter would use about 7.3 global ha (18.1 global acres) (Global Footprint Network 2009). Some ecological footprint calculators can also be used to measure dietary ecological footprints in isolation from other components. The Earth Day Network (2008), an organisation based in the USA to promote, together with an international network of partners, their belief in a moral human right to a healthy and sustainable environment, has developed such a calculator. The calculator can be used, for example, to measure the dietary ecological footprint of a male person, aged between 36 and 50, living not far from Edinburgh (Scotland), who eats very few foods that are processed, packaged, or imported. To satisfy his dietary needs, such a person would require 0.5 global ha if he never eats animal products, 0.9 global ha if he eats such products occasionally (defined as: no meat or occasionally meat, but eggs/dairy almost daily), or 1.4 global ha if he eats meat very often (defined as: daily). Whereas these findings cannot be generalised across different countries because of ecosystemic and technological differences, several studies carried out in various countries have concluded that the ecological footprints of many diets that include farmed animal products exceed those of many other diets (see, e.g. Baroni et al. 2007; Carlsson-Kanyama and González 2009; Eshel and Martin 2009; Peters, Wilkins, and Fick 2007; Reijnders and Soret 2003).

While the main focus of this article is on farmed animal products, it is worth mentioning that many products derived from animals who have not been farmed do not fare much better. In this respect, the ecological footprints associated with the consumption of fish are particularly worth mentioning. The argument has been made that the costs related to the environmental degradation (and species losses) caused by (over)fishing are significant and routinely undervalued (Myers and Worm 2003; Worm et al. 2006). Having examined the energy costs of fishing, Eshel and Martin (2006) conclude that typical Western diets that include fish are more inefficient when compared with plant-based diets, especially since long-distance boat journeys are associated with the catching of fish preferred by Western customers. This was confirmed by Reijnders and Soret (2003, 667S), who claim that the fossil fuel requirements for trawler fishing – the prevailing fishing method in Western Europe – is 14 times larger than the amount needed to produce an equal amount of plant protein in Western Europe. This excludes the high emissions that are frequently used to process fish, for example the emissions produced by canning and refrigeration. High emissions for cod fishing were also calculated by Carlsson-Kanyama and González (2009, 1707S). Whereas many aspects of the fishing industry are associated with relatively large ecological footprints, it must be pointed out that the consumption of some fish (e.g. herbivorous fish kept in nearby ponds) can be associated with relatively small ecological footprints, but their potential to contribute to a substantial part of the diets of the growing human population may be limited due to space, time, and ethical constraints. Whereas some methods to farm fish are associated with relatively small footprints, other methods used in aquaculture have been associated with relatively large ecological footprints because of their use of pesticides, prophylactic antibiotics, eutrophying nutrients, and their use of other fish as feed (Cole et al. 2009). With regard to this last factor, one study estimates that about one-third of all the fish who are caught is used to feed farm animals (Goldburg and Naylor 2005, 23). A different study, however, estimates that the aquaculture industry used as much as 68.2% of all fish meal and 88.5% of all fish oil that were produced from the bodies of fish in 2006 (Tacon and Metian 2008).

Whereas the ecological footprint is a more inclusive measure than that provided by greenhouse gas emissions alone, it was not designed to be a tool to measure the moral character of human actions. This is so for good reasons. While some activities may use relatively few resources and produce little waste, they might nevertheless undermine human and nonhuman health. And vice versa, a relatively large ecological footprint need not necessarily be associated with relatively large negative GHIs, for example, in situations where activities that have relatively large ecological footprints produce relatively large health benefits. Whereas it may be more difficult to quantify GHIs as people may have widely differing ideas about what should count as a health impact and about how they should be measured, this shows why we need such a measure. It is impossible to say categorically that those who produce relatively large amounts of greenhouse gas emissions or those who have relatively large ecological footprints act immorally. In order to evaluate the morality of their actions with regard to their impacts on human beings, we must examine whether the negative GHIs that are produced by their actions violate any human being's right to health protection.

In spite of their limited use to evaluate the moral character of human actions, those who have relatively large ecological footprints would have relatively large negative GHIs unless they produce relatively large quantities of positive GHIs by their actions. It must be concluded, if Rees (2006) is correct, that the ecological footprint of the global human population today jeopardises the availability of sufficient resources to meet the basic health needs of those who will be living in the future. Whereas some of the negative GHIs associated with the production systems that underlie omnivorous diets might be offset by the positive GHIs that they produce (including potential nonfood-related benefits), it must be questioned whether diets that exclude farmed animal products are more likely to protect the right to health protection than diets that include them. Whereas people may have widely different views about the scope of such a right (i.e. in relation to the question of what health is, and whether a distinction between basic health protection and nonbasic health protection should be made), about what should count as a negative and a positive GHI, and about how to measure the relative merits and demerits of different diets, the analysis presented above suggests that the answer to this question may be positive for people living in a wide range of different locations. Whereas a more developed theory would include an account of how negative and positive GHIs might be quantified, this need not exclude the possibility of taking into consideration those negative and positive GHIs that are difficult to quantify, say, for example, the amount of pleasure people derive from eating particular foods, the degree of importance they give to particular risks and uncertainties (e.g. related to zoonotic diseases), the benefits that some people derive from the traction power or the aesthetic value that some animals may provide, or the need to consider deontological constraints apart from the human right to health protection, for example in relation to the question of whether any nonhuman animals should be given a similar right.

## Why some people might need to curtail their consumption of farmed animal products

If the analysis presented above is plausible, many people could reduce their negative GHIs by reducing their intakes of farmed animal products. Therefore, if people who exceed their fair share of negative GHIs are morally obliged to reduce them, they might consider making dietary changes. In this light, a more radical proposal than McMichael's has been put forward by Compassion in World Farming (2007, par. 6.1), who claim that those living in 'developed countries should reduce production and consumption of meat and milk to at least 60% below current levels by 2050'. My view is that this should be a moral obligation only for those people (whichever country they live in) who would exceed their fair share of negative GHIs if they failed to do so and who prefer this method of reducing their negative GHIs over any other methods that might be available.

However, when we compare the potential of reducing or eliminating farmed animal products from our diets with some alternative methods that we could adopt to reduce our negative GHIs, two striking differences can be observed. First, while it might take many years before we have developed and implemented more sustainable energy and transportation technologies, dietary changes can be made right now. We need not await the arrival and implementation of new technologies. This is highly relevant since people who exceed their fair share of negative GHIs are in debt right now, and therefore must act immediately to reduce their negative GHIs. A second difference is that, whereas a reduction in the intake of animal products is associated with health benefits for many people who can obtain adequate alternative foods, significant costs are associated with most other methods we have at our disposal to reduce our negative GHIs at the present time.

To illustrate the latter, I shall give four examples from sectors that contribute significantly to people's negative GHIs. First, if we cannot heat our buildings without making significant cuts in our use of resources, we shall have to lower our heating requirements if we want to reduce our consumption in this sector. For those who have them, the use of a range of electrical appliances, for example, washing machines, kettles, lighting, audiovisual equipment, and computers could also be curtailed. A third area where we could reduce our energy consumption is in the purchase of nonfood items, especially items that require a lot of resources and energy in their life-cycles. A fourth sector that dissipates a lot of energy is the transport sector. We could reduce the transportation of goods that do not satisfy vital interests and limit private transportation, for example, by limiting air and car travel. Alternatively, people could use cleaner fuels to reduce their greenhouse gas emissions. However, whereas reducing greenhouse gas emissions *per se* would be positive, the cleaner fuels that replace them might not necessarily decrease negative GHIs. For example, even if biofuels were grown without the use of chemicals (such as synthetic fertilisers and pesticides) produced from fossil fuels, their production might still incur relatively large amounts of negative GHIs. The land on which biofuels can be grown could be used for other purposes that might produce relatively fewer negative GHIs per unit of benefit.

Whereas Caney (2009, 179) has claimed that cuts in greenhouse gas emissions 'from driving cars, taking plane flights, poorly insulated housing, and inefficient energy use ... cannot be said to compromise any human rights', no leap of imagination is required to suspect that for many people who exceed their fair share of negative GHIs, it may be very demanding to make significant cuts in these areas, as well as more generally in any of the four sectors I have mentioned. If this is so, they might agree with the view that they are under a moral obligation to limit their negative GHIs to what is their fair share by restricting the consumption of farmed animal products where doing so would not compromise their ability to access adequate nutrition. Doing so might be far less costly and provide not only social but also individual and environmental gains (Eshel and Martin 2009, 1713S).

## Conclusion

Calculations of greenhouse gas emissions and ecological footprint calculations reveal that in most contexts, diets that include the consumption of farmed animal products produce more greenhouse gas emissions and have larger ecological footprints, compared to other diets. Whereas these calculations are important, it has been argued that they must be seen as important aids for decision-making, rather than decision-making tools themselves. More specifically, I have argued that, in order to evaluate dietary choices, these choices must not be treated in isolation from the other choices we make. While obtaining accurate information about the greenhouse gas emissions and ecological footprints of different dietary options is important, this knowledge does not tell us anything about why some diets might be more problematic than others.

To answer this question, I applied an incomplete theory of human justice that focuses on the negative duty we have to make sure that our actions do not jeopardise the health of other humans unfairly. In order to measure how human actions influence health, I applied the 'Global Health Impact' concept, a novel unit of measurement that I introduced in a previous work (Deckers 2010b, 2011b). I argued that diets that include farmed animal products are – in many situations – associated with more negative GHIs when compared with alternative diets. While many different strategies could be adopted by those who are under a moral obligation to constrain their negative GHIs, I have suggested that curtailing the consumption of farmed animal products might be an option that is less demanding and even beneficial, compared with the alternative options that may be available.

In this article, I have ignored the question of what the negative GHIs of different human diets are on the organisms that make up the nonhuman world. I have argued elsewhere, however, that we should also consider the health impacts upon the nonhuman world when we determine what counts as a fair share of negative GHIs, and that – in many situations – the negative GHIs on the nonhuman world that are associated with diets that include farmed animal products exceed those that are associated with diets that exclude them (Deckers 2011c). If the argument developed there survives ethical scrutiny, it adds further weight to the argument that in many situations, the negative GHIs of the former are considerably greater than those of the latter.

## References

Aarestrup, F., Y. Agerso, P. Gerner-Smidt, M. Madsen, and L. Jensen. 2000. Comparison of antimicrobial resistance phenotypes and resistance genes in *Enterococcus faecalis* and *Enterococcus faecium* from humans in the community, broilers, and pigs in Denmark. *Diagnostic Microbiology and Infectious Disease* 37, no. 2: 127–37.

Anomaly, J. 2010. Combating resistance: The case for a global antibiotics treaty. *Public Health Ethics* 3, no. 1: 13–22.

Baroni, L., L. Cenci, M. Tettamanti, and M. Berati. 2007. Evaluating the environmental impact of various dietary patterns combined with different food production systems. *European Journal of Clinical Nutrition* 61, no. 2: 279–86.

Bell, D. 2011. Does anthropogenic climate change violate human rights? *Critical Review of International Social and Political Philosophy* 14, no. 2: 99–124.

Belshe, R. 2005. The origins of pandemic influenza – Lessons from the 1918 virus. *New England Journal of Medicine* 353, no. 21: 2209–11.

Burkholder, J., B. Libra, P. Weyer, S. Heathcote, D. Kolpin, P. Thorne, and M. Wichman. 2007. Impacts of waste from concentrated animal feeding operations on water quality. *Environmental Health Perspectives* 115, no. 2: 308–12.

Caney, S. 2009. Climate change and the future: Discounting for time, wealth, and risk. *Journal of Social Philosophy* 40, no. 2: 163–86.

Carlsson-Kanyama, A., and A. González. 2009. Potential contributions of food consumption patterns to climate change. *American Journal of Clinical Nutrition* 89 (Suppl.): 1704S–9S.

Chapman, H., and Z. Johnson. 2002. Use of antibiotics and roxarsone in broiler chickens in the USA: Analysis for the years 1995 to 2000. *Poultry Science* 81, no. 3: 356–64.

Cole, D., R. Cole, S. Gaydos, J. Gray, G. Hyland, M. Jacques, N. Powell-Dunford, C. Sawhney, and W.W. Au. 2009. Aquaculture: Environmental, toxicological, and health issues. *International Journal of Hygiene and Environmental Health* 212, no. 4: 369–77.

Collinge, J. 1999. Variant Creutzfeldt–Jakob disease. *Lancet* 354, no. 9175: 317–23.

Compassion in World Farming. 2007. *Global warning: Climate change and farm animal welfare*. Godalming: Compassion in World Farming.

Deckers, J. 2010a. Should the consumption of farmed animal products be restricted, and if so, by how much? *Food Policy* 35, no. 6: 497–503.

Deckers, J. 2010b. What policy should be adopted to curtail the negative GHIs associated with the consumption of farmed animal products? *Res Publica* 16, no. 1: 57–72.

Deckers, J. 2011a. Could some people be wronged by contracting swine flu? A case discussion on the links between the farm animal sector and human disease. *Journal of Medical Ethics* 37, no. 6: 354–56.

Deckers, J. 2011b. Negative 'GHIs', the right to health protection, and future generations. *Journal of Bioethical Inquiry* 8, no. 2: 165–76.

Deckers, J. 2011c. Should Whiteheadians be vegetarians? A critical analysis of the thoughts of Hartshorne and Dombrowski. *Journal of Animal Ethics* 1, no. 2: 195–209.

Duffy, G., and E. Moriarty. 2003. Cryptosporidium and its potential as a food-borne pathogen. *Animal Health Research Reviews* 4, no. 2: 95–107.

Earth Day Network. 2008. Ecological footprint quiz. http://www.earthday.net/footprint/.

Eik, L., G. Kifaro, S. Kiango, Ø. Nordhagen, J. Safari, and L. Mtenga. 2008. Productivity of goats and their contribution to household food security in high potential areas of East Africa: A case of Mgeta, Tanzania. *African Journal of Food, Agriculture, Nutrition and Development* 8, no. 3: 278–90.

Eshel, G., and P. Martin. 2006. Diet, energy, and global warming. *Earth Interactions* 10, no. 9: 1–17.

Eshel, G., and P. Martin. 2009. Geophysics and nutritional science: Toward a novel, unified paradigm. *American Journal of Clinical Nutrition* 89 (Suppl.): 1710S–16S.

Fraser, C., C. Donnelly, S. Cauchemez, W. Hanage, M. Van Kerkhove, and T. Hollingsworth. 2009. Pandemic potential of a strain of Influenza A (H1N1): Early findings. *Science* 324, no. 5934: 1557–61.

Garnett, T. 2009. Livestock-related greenhouse gas emissions: Impacts and options for policy makers. *Environmental Science and Policy* 12, no. 4: 491–503.

Global Footprint Network. 2009. http://www.footprintnetwork.org/en/index.php/GFN/page/calculators/.

Goldburg, R., and R. Naylor. 2005. Future seascapes, fishing, and fish farming. *Frontiers in Ecology and the Environment* 3, no. 1: 21–8.

Goodland, R., and J. Anhang. 2009. Livestock and climate change. What if the key actors in climate change are ... cows, pigs, and chickens? *WorldWatch*, November/December: 10–19.

Greger, M. 2007. The human/animal interface: Emergence and resurgence of zoonotic infectious diseases. *Critical Reviews in Microbiology* 33, no. 4: 243–99.

Herrero, M., and P. Thornton. 2010. Mixed crop livestock systems in the developing world: Present and future. *Advances in Animal Biosciences* 1, no. 2: 481–2.

Intergovernmental Panel on Climate Change (IPCC). 2007. *Climate change 2007: Synthesis report. Summary for policymakers*. http://www.ipcc.ch/pdf/assessment-report/ar4/syr/ar4_syr_spm.pdf.

Leroy, J., and E. Frongillo. 2007. Can intervention to promote animal production ameliorate undernutrition? *Journal of Nutrition* 137, no. 10: 2311–16.

Lloyd-Williams, F., M. O'Flaherty, M. Mwatsama, C. Birt, R. Ireland, and S. Capewell. 2008. Estimating the cardiovascular mortality burden attributable to the European Common Agricultural Policy on dietary saturated fats. *Bulletin of the World Health Organization* 86, no. 7: 535–41.

McMichael, A., J. Powles, C. Butler, and R. Uauy. 2007. Food, livestock production, energy, climate change, and health. *Lancet* 370, no. 9594: 1253–63.

Messina, V., and K. Burke. 1997. Position of the American Dietetic Association: Vegetarian diets. *Journal of the American Dietetic Association* 97, no. 11: 1317–21.

Müller, O., and M. Krawinkel. 2005. Malnutrition and health in developing countries. *Canadian Medical Association Journal* 173, no. 3: 279–86.

Myers, R., and B. Worm. 2003. Rapid worldwide depletion of predatory fish communities. *Nature* 423, no. 6937: 280–3.

National Academy of Sciences. 2006. *Health risks from dioxin and related compounds*. Washington, DC: National Academies Press.

Pagnamenta, R. 2009. Climate Chief Lord Stern: Give up meat to save the planet. *The Times*, October 27. http://www.timesonline.co.uk/tol/news/environment/article6891362.ece

Peters, C., J. Wilkins, and G. Fick. 2007. Testing a complete-diet model for estimating the land resource requirements of food consumption and agricultural carrying capacity: The New York State example. *Renewable Agriculture and Food Systems* 22, no. 2: 145–53.

Rees, W. 2006. Ecological footprints and bio-capacity: Essential elements in sustainability assessment. In *Renewables-based technology: Sustainability assessment*, ed. J. Dewulf and H. Van Langenhove, 143–58. Chichester: John Wiley & Sons.

Reijnders, L., and S. Soret. 2003. Quantification of the environmental impact of different dietary protein choices. *American Journal of Clinical Nutrition* 78, no. 3: 664S–8S.

Rosales, J. 2008. Economic growth, climate change, biodiversity loss: Distributive justice for the global north and south. *Conservation Biology* 22, no. 6: 1409–17.

Sanders, T. 1999. Meat or wheat for the new millennium? A debate. Pro veg. The nutritional adequacy of plant-based diets. *Proceedings of the Nutrition Society* 58, no. 2: 265–9.

Sapkota, A., L. Lefferts, S. McKenzie, and P. Walker. 2007. What do we feed to food-production animals? A review of animal feed ingredients and their potential impacts on human health. *Environmental Health Perspectives* 115, no. 5: 663–70.

Stahler, C. 2006. How many adults are vegetarian? The Vegetarian Resource Group asked in a 2006 national poll. *Vegetarian Journal* 25, no. 4: 42–4.

Steinfeld, H., P. Gerber, T. Wassenaar, V. Castel, M. Rosales, and C. de Haan. 2006. *Livestock's long shadow. Environmental issues and options*. Rome: Food and Agriculture Organization of the United Nations.

Stern, N. 2006. *The economics of climate change*. Cambridge: Cambridge University Press.

Swedenborg, E., J. Ruegg, S. Makela, and I. Pongratz. 2009. Endocrine disruptive chemicals: Mechanisms of action and involvement in metabolic disorders. *Journal of Molecular Endocrinology* 43, no. 1: 1–10.

Tacon, A., and M. Metian. 2008. Global overview on the use of fish meal and fish oil in industrially compounded aquafeeds: Trends and future prospects. *Aquaculture* 285, nos. 1–4: 146–58.

Trasande, L., P. Landrigan, and C. Schechter. 2005. Public health and economic consequences of methyl mercury toxicity to the developing brain. *Environmental Health* Perspectives 113, no. 5: 590–6.

UK Parliament. 2008. *Climate Change Act 2008*. London: The Stationery Office Limited.

Wackernagel, M., and W. Rees. 1996. *Our ecological footprint: Reducing human impact on the earth.* Gabriola Island, BC: New Society Publishers.

Weidema, B., M. Wesnaes, J. Hermansen, T. Kristensen, and N. Halberg. 2009. *Environmental improvement potentials of meat and dairy products.* Luxembourg: Office for Official Publications of the European Communities.

WHO (World Health Organization). 2003. *Diet, nutrition, and the prevention of chronic diseases. WHO Technical Report Series No. 916.* Geneva: World Health Organization.

Woolhouse, M., and S. Gowtage-Sequeria. 2005. Host range and emerging and reemerging pathogens. *Emerging Infectious Diseases* 11, no. 12: 1842–47.

World Cancer Research Fund and American Institute for Cancer Research. 2007. *Food, nutrition, physical activity, and the prevention of cancer: A global perspective.* Washington, DC: American Institute for Cancer Research.

Worm, B., E. Barbier, N. Beaumont, J. Duffy, C. Folke, B. Halpern, J. Jackson, et al. 2006. Impacts of biodiversity loss on ocean ecosystem services. *Science* 314, no. 5800: 787–90.

# Climate ethics: The end of development or a new paradigm through African sagacity?

Sirkku Hellsten[a], Frederick Ochieng'-Odhiambo[b] and Martin Schönfeld[c]

[a]*University of Helsinki/University of Dar es Salaam;* [b]*University of the West Indies;* [c]*University of South Florida*

## 1. Introduction

By this time and age there has been worldwide recognition that climate change, development, human wellbeing, and security all go hand-in-hand. The World Commission on Environment and Development (WCED), the United Nations Intergovernmental Panel on Climate Change (IPCC), and the UN High Level Panel on Global Sustainability (GSP) have noted clearly that the Earth cannot sustain development based on ever increasing economic growth and consumption that causes environmental degradation, resource depletion, and climate change.[1] Similar reports have been issued by the Worldwatch Institute and other NGOs. There has been much research on the impact and effects of climate change in different parts in the world (see Schönfeld 2011, 129-136, Schussler 2011; see also endnote 4).

Global and national development strategies are finally acknowledging the connections of development, wellbeing, and security to climate change. But the concepts to tie these interdependencies together, most prominently 'sustainable development' and 'green economy', retain the conventional growth-based model of development. Economic growth evidently requires more consumption, more industrialization, and more resource use. The need for growth is based on our expectations of a higher standard of living.[2]

Despite global concern and increased awareness of climate change and its consequences, little has been achieved so far. There is a scientific consensus, but there is no policy consensus on the issues. Making binding international agreements has turned out to be next to impossible despite the high hopes set on the Kyoto Protocol and its follow-up Conferences of the Parties (e.g., COP-15 Copenhagen 2009, COP-16 Cancún 2010, COP-17 Durban 2011, and, in 2012, Rio +20). The lack of achievement is chiefly due to the obstruction by the world's biggest polluters – the USA and China.[3]

Among the efforts to adapt to the consequences of climate change are national plans to reduce the vulnerability of their ecosystems and communities. This is important but also shows a continued commitment to the growth-based concept of development. The adaptive efforts seem to be little more than proactive damage control. They also suggest an unfounded optimism, that with the help of green technologies, production systems, and consumer goods, we can find a way out of the mess we have made. We hope that despite the evident and radical changes happening around us, we can keep on the current path, we can improve living standards, and we will not have to give up the luxuries we are used to. The dual crises of climate change and of environmental degradation are seen as problems that require an adjustment and fine-tuning of conventional economic activities instead of a more fundamental transformation of our civilization, lifestyles, and philosophies.

All in all, besides feeble acknowledgements that 'somebody should do something' about climate change, there is little indication that we will see any time soon a revision of the standard model of development. The international community is not ready to abandon a model that the Western world has persistently exported to other cultures – those considered 'less developed', 'under-developed', and 'uncivilized' not that long ago. Criticism is mounting, but the paradigm change that Martin Schönfeld has called for (Schönfeld 2011: 129 – 136) is still wanting, though urgently needed.

This chapter assumes the interconnectedness of development, nature, well-being and security. The authors argue that we need to craft a new conception of development if we are to deal with climate change and its consequences. This alternative conception is based on a holistic framework. This means to shift our understanding of humans from masters of nature to members of nature, who are placed in a natural order that controls us all. The authors do not aim to disregard contractarian and rights-based approaches, but they see their defects, and thus argue for the need to focus more on our responsibilities and duties towards nature, even to the universe that has given us life. In a sense, 'climate' concerns a wider discourse than that of environment; in another sense the universe gives us a certain climate to live by and live within. A holistic understanding contains the realization that we neither exist as independent atoms nor live in an atomistic world, and that everything, including humans, is interrelated and belongs together. We recognize the rights of individuals from a stance that gives weight to corresponding duties and to local and global responsibilities. We argue that effective responses to climate change require a shift of our cultural, social, and economic standards, and of their implicit metaphysics and ethics, to a communalist way of thinking.

The idea of communalism draws its foundation from African philosophy and particularly sagacity. Sagacity is the old wisdom of those who knew how our species (ancestors, the living, and future generations) could live with itself and in harmony with nature, as unforgiving as nature could at times be. This is not so much the specific wisdom of certain thinkers but rather a more collectivist approach to the world. African sages had to live in a climate that could be harsh, and they had to learn the ways of nature – what the community should give to it, and what it can give to the community. Sagacity is not knowledge in a modern, scientific sense. It is the collective wisdom of those whose descendants will suffer most from climate change and the consequent toughness of living conditions. We want to borrow from the holistic thinking of those who live in regions most vulnerable to climatic changes.

This book contains chapters that also cite the key findings of climate science (that climate change is happening, that it is perpetrated by humans, that it is accelerating, and that it is bad for our prospects). We proceed from the same starting point, and if necessary, refer to other essays in the collection. Since the other chapters examine Eastern, Western, and Northern perspectives on climate ethics, we shall explore the Southern, specifically African, contribution to climate ethics. This contribution is often neglected, even though Africa is a continent very much in the middle of the climate debate, by way of development policies, resource management, and poverty reduction.[4] The effects of climate change make themselves felt in various African regions and worsen the plight of the poorest and most vulnerable Africans. Africa is also a place whose middle class embraces Western consumerism and where corrupt governments sell out the last remaining resources of their lands to multinational corporations (which, for their part, ensure that the faith in economic growth remains unassailable). We want to examine whether traditional African philosophy can offer new ideas to take climate ethics forward.

## 2. The Western concept of development & climate change: Can we sustain more development?

Since the onset of modernity, the West has advanced a model of development that identifies development with progress and modernization, and more recently, with wealth, purchasing power, and consumer choices. This model has been set up as an ideal that the rest of the 'developing' or 'less developed' world ought to strive for. First through colonialization (which colonial powers used to defend as modernization projects and as 'civilization missions'), and later through global trade and economic liberalism, the West has exported its own unsustainable lifestyle to the rest of the world. But now, that the unsustainability of the Western model is evident for all to see, global civilization finds itself at a crossroads. The choice is not between 'sustainable' and 'unsustainable' growth, but rather between growth and no growth, and thus a choice that concerns the basic assumptions of our civilization. The globalized liberal economies have marginalized ideological and cultural alternatives; market liberalism has effectively conquered the planet. But we need to keep in mind that no conquest is forever, and that other civilizations before this one have fallen. (For further criticism of the Western concept of development see Escobar 1985, Janz 2011, 173 – 184, Rist 2002, Sachs 1992, Schönfeld 2011, 129 – 136, Schuurman 2000, Subramaniam et al. 2002).

Given the gravity of the consequences of climate change, and the inability of the growth-based model of development to offer long-term solutions, there is a need for a paradigm change regarding the direction which human development should take. But global capitalism is so intertwined with the 'development industry', that it is difficult to let go of the belief in the possibility of endless growth. It also seems difficult to own up to the presumption that to keep growth going, we need more people to consume more. The burgeoning middle class in the BRIC-nations (Brazil, Russia, India, and China) and in many African countries shows how much conventional development keeps affecting climate, habitats, and cultures.[5]

It is ironic that Western capitalism is now languishing in crisis just when the rising economies of Asia and Africa want a bigger share of the cake that Western businesses used to have all for themselves. Not long ago, Western pundits were praising the 'Asian tigers' for their performance and were happy about the creation of these new markets.

Today, Asian and also African economies are expanding consumption, are enlarging their monetary reserves, and are beginning to dominate world markets. Some of the poorest and most stratified countries in the world, such as Mozambique and Angola, enjoy at present the highest economic growth rates (over seven percent), while growth in the U.S. and the E.U. is flagging.

But while 'less developed' countries are gaining in economic strength, other trends are not keeping up. For example, progress towards development goals – such as full employment, good governance, economic justice, public health, the eradication of poverty, and the reduction of birth rates – whose realization is crucial to wellbeing and security has not kept pace with their anticipated economic development or strength. Furthermore, economic development not only widens social stratification, but also undermines environmental integrity. Not only in the Western countries but also in Africa, elites become ever richer, and maintain large carbon footprints and indulge in lifestyles that are environmentally hazardous. But as the poor try to survive, they may have to do so in ways that are environmentally hazardous as well. The problem is not the traditional conflict between the rich North and the poor South anymore. The problem has changed into a global conflict between predatory capitalism spawning wealthy oligarchies, and powerless masses locked into poverty. The solution will require a concerted effort to work towards a concept of development that seriously takes in account those now marginalized by economic development.

## 3. Climate change: Ethics, rhetoric and politics

Governments and intergovernmental organizations discuss funding mechanism for mitigating climate change such as carbon taxes on primary emitters, or loans and grants to poor countries. Such discussions require delegates to fly across continents (leaving behind large carbon footprints), without having led to tangible results: if there are agreements, they are hardly ever adopted, and if they are adopted, they hardly ever become binding. Meanwhile, Western countries support development programs meant to funnel more wealth to poor countries, while new economic agents such as the BRIC are opening new markets. The gap between rich and poor is not narrowing, and environmental damages are not lessening. New discoveries of oil fields in Africa bring more consumption and pollution, and encourage the fossil fuel giants to continue with business as usual, as if peak oil has not yet arrived. That there are still many opportunities of mining and exploiting traditional energy resources does not help the switchover to renewables, postcarbon technology, and sustainable commodities. We may well be headed towards the end of development if we keep going in this direction. This outcome can be avoided by recasting the flourishing of existence in ways other than materialistic wealth creation. But it is a daunting challenge to realize that the Western model of endless, linear, and materialistic development is nothing but a myth.

Among the attempts to account for alternative measures of development, there is the *World Happiness Report*, edited by J. Helliwell et al. (2012, and the publications by A. Sen and M. Nussbaum on the 'capability approach' to development (1984 and 1987). Such works emphasize that the concept of development is not exhausted by notions such as GNP or yardsticks such as economic growth. Bigger is not always better. Development also contains the dimensions of freedom and wellbeing. It is also noteworthy that there is little correlation between wealth and happiness – people in affluent countries, with the easiest access to consumer goods, are by no means happier than people in poorer countries; neither are they more functional with regard to the development of human capabilities. While these attempts did have some impact on development indicators, by yielding multidimensional models of poverty and wellbeing, they failed to change the direction of development-models. Their sole lasting influence, to date, has been to make the development agenda more complex; an agenda that now also aims for 'inclusive growth', 'green economies', and 'sustainable growth'.

The preservation of the sanctity of the growth paradigm is motivated by the traditional Western view of reality, with its atomistic, individualistic, and materialistic aspects, and which applies development so conceived to culture, economy, and technology. In this view, development is a continuous and linear process, from worse to better, or from lower to higher state of affairs. The collision with the biospherical limits of growth, whose consequence is climate crisis, discloses the inherent flaw of this view.

The unintended collision with the limits of growth indicates the lack of realism in the linear-progressive model of development. If there is anything science has taught us, then it is that there are no straight lines in the universe; rather, there are beginnings and cumulative expansions and sudden implosions and extinctions. Development, in a natural sense, is always a cycle; it waxes and wanes, it begins and ends. Already in this regard, there is something oddly unnatural and unreal about the notion of linear growth.

Development, in a natural sense, also reflects the interconnectedness of all parts affected by it. In nature, development 'works out' if all affected parts grow in a collective, holistic balance. Organic and ecologic examples come to mind: functional development occurs when all structures and tissues in a growing body develop in equal measure, or when a thriving ecosystem enhances its overall complexity over time until it matures into a climax community such as a coral reef or a rainforest. By contrast, development of a single component or aspect tends to be destructive to the whole. Such non-holistic, dysfunctional development occurs with a neoplasm metastasizing and overrunning a body with cancer, or with a blight taking over an ecosystem, crashing its biodiversity and reducing its complexity, transforming coral beds into sea grass or a rainforest into a steppe.

If we take our cue from the holistic structure of functional natural development, then linear economic growth needs to be replaced by a more comprehensive framework for development, such as wellbeing, and must include the development of social justice, political participation, and economic as well as social equality.[6] Last but not least, we must remember that this is not a purely theoretical point. Although climate change is now irreversible, there is still a window of opportunity for action to ensure it will not escalate. But seizing this chance requires abandoning the self-destructive ways of living that are intertwined with linear economic growth. Thus for practical reasons as well, we need to create a new development paradigm, based on the needs of community and humanity, the 'community of all communities', in the words of the Tanzanian philosopher and first president Julius Nyerere (1968).

## 4. What are the alternatives to the conventional development model?

In the context of ethical responses to climate change, M. Schönfeld sees reasons for hope in new approaches in environmental and global ethics. But the problem with these attempts is that they suggest long-term action plans, which may not be suited to the urgency and speed of the global transformations. International institutions, intergovernmental organizations, and global movements are realizing the vastness and immediacy of the climate crisis. That climate-related panels, commissions, and inquiries are multiplying now is a step in the right direction. But we need to move from rhetoric to politics and from politics to practice. The realization that the Western model of development is flawed does not mean we should not try to work towards progress – to make the world a better place, to seek a better quality of life, and to enhance social justice and economic equality.

The post-development critique that draws from the Latin American dependency theory, and the post-colonial critique that has roots in African liberation from colonial rule, converge in the identification of central problems in the Western concept of development. Some of these criticisms draw from Foucault's concept of bio-politics and its deconstruction of social power relations (Escobar 1985, Esteva 1992, Janz 2011, 173 – 184, Rist 2002, Sachs 1992, Schuurman 2000, Subramaniam et al. 2002). The most recent 'State of the World' report of Worldwatch Institute, *Moving towards sustainable prosperity* (2012), goes further; its authors argue for a reduction and reversal of growth in countries they consider overdeveloped. Overdeveloped countries are those whose consumerism destroys the changes for equitable and sustainable development. Such countries must assume a new duty, the obligation of *degrowth* (Assadourian 2012). The concept of degrowth underscores an emerging consensus in the environmental sciences that economic growth and biospherical integrity are incompatible, despite former hopes to the contrary. The alternative to growth would be sustainable prosperity for all, which takes into account existential wellbeing instead of consumer agency. We need global cooperation and more revisionist ideas to flesh out this conceptual alternative and implement it as a future-oriented policy. The question that deserves serious consideration is this: what kind of life is actually worth striving for? A derivative question would be: what will a truly cosmopolitan solution to the climate crisis look like?

One of the main problems of Western ethics has been the preponderance of contractarian and rights-based approaches (see Pogge 2002, Löfquist 2011, Schussler 2011). While the recognition of individual rights is important, the evolution of ethics cannot proceed without a corresponding emphasis on complementary duties. As the resounding failure of twenty years of climate politics since the first Earth Summit at Rio illustrates, it is not sufficient to rely on contractarian, rights-based approaches, since there is no way to enforce any such intergovernmental agreements and treaties. In the end, governments, organizations, and enterprises are left to act in accordance with their own conscience. This is no longer enough.

The rights-based approach to justice and order has created a quasi-Hobbesian world, in which everyone seems to have a right to everything but where there is simply not enough to share with all. The 'right to development', in particular, has become a problematic concept, and as we all claim rights, we cannot deny them to others. But, as Bentham reminds us, there are no rights without corresponding duties. So we must determine which claimed rights impose justifiable duties, and what specific responsibilities they would entail. Switching our perspective from rights to duties, and from entitlements to responsibilities, brings the African approach into focus. This approach is a communalist model, and it suggests itself as a global ethics if one expands the domain of its key notion, solidarity, from neighbors in a local community to the network of all communities constitutive of the human society. In the age of climate change, the normative aspiration of this model would be to minimize consumption by maximizing enjoyment – to make do with less, while getting more out of life.

## 5. The African perspective: Communalism and personhood in African cultures

Until now, the African approach has received little attention in the context of climate ethics. Our starting point is the claim that the climate crisis cannot be meaningfully analyzed without the interrogation of the aims and direction of development. While such an interrogation is gaining momentum as part of a global development ethics movement, its discursive framework remains that of the Western model. Abandoning this framework requires what post-development and post-colonial critique calls 'decolonization of the mind'. We believe that such a decolonization of the mind is one of the most important turns we need to make. This heuristic process should not be restricted to Africa or to the 'developing' nations as such; it should also encompass the affluent nations of the North, which, as primary international decision-makers, they have an even greater need for this process.

Of course, changing the course of civilization to combat climate change will not be easy. But it is our conviction that the transformation to a postcarbon, sustainable, and resilient world cannot be completed by adhering to the individualist paradigm of international ethics, politics, or economics – no matter what business representatives, political elites, and development agencies might claim. Neither do we think that the quest among traditional global ethicists for new 'cosmopolitan' perspectives about human development will lead in the desired direction. What is needed, instead, is a discursive shift towards justice, responsibilities, solidarity, and sharing which still respects the lives and rights of the individuals within their community and within the community of communities (Nyerere 1968). It should be clarified here that we are critical towards the Western cosmopolitanism that still has its foundations in liberal individualism and its concept of tolerance extend only as far as the different cultures are ready to assimilate liberal values. The communitarian view promoted in this chapter is not opposed to cosmopolitanism as world citizenship but is ontologically different. The difference is that it is based on African philosophy which sees individuals and communities not only as interdependent but also acknowledges that our communities are not 'autonomous agents'. Its basic tenet is that we are all part of various communities which overlap, and in the end expand to a community that covers the whole earth (see Appiah 2006, Nyerere 1968, see also Hellsten 2011).

There is abundant literature about the communalistic spirit of traditional African societies; for example by Senghor (1964), Nyerere (1968), Tempels (1969), van der Walt (1997), and Ikuenobe (2006). We do not wish to burden the reader with the details of that social system. It is sufficient for our purposes to give a synopsis of this communalism, given that its key elements have a cardinal role to play in climate ethics.

South African theologian and philosopher Bennie van der Walt maps out the ontological, anthropological, epistemic, and axiological distinctions between African and Western societies; that is, their divergent conceptions of reality, humans, knowledge, and values. He shows that the most prominent difference is that the African culture prioritizes community over individuals, and that its modes of thinking are inseparable from the communal framework. Further cultural differences include primacy being given to spiritual powers in Africa, whilst in the West currency is given to material entities; Africans see themselves as close to, or part of concrete reality, while Westerners discuss reality in abstract terms as if they were not part of it; Africans tend to get involved with objects of knowledge, while Westerners observe objects of knowledge at a distance; and generally, African modes of thinking are less analytical, more synthetic, and rather intuitive than reflective, while Westerners are the opposite (see van der Walt 1997, 81 – 82).

At this juncture one can see how not only Western modes of production but also Western mentalities are responsible for the climate crisis; why the international community, still largely led by Western powers, has failed to mitigate and rein in climate change; and also why, in global ethics, the materialistic-linear paradigm of development prevails. Human society is largely guided by Western *weltanschauung*, and in this worldview materialism constitutes the yardstick for measuring quality of life and existential success, and is also a determining factor in ethical choices and aesthetic deliberations. It certainly does not help that, in this *weltanschauung*, individuals see themselves as essentially separate from reality, and relate to this reality primarily in objectifying and utilitarian ways, consistently serving only their own interests.[7] And this is "where and when the rain begins to beat us." In a predominantly spiritual worldview, by contrast, in which individuals see themselves as integral parts of reality, wanton destruction of nature would only occur as a remote unconscious possibility. Building on the African approach for coming to terms with the climate crisis is accordingly a worthwhile effort.

In his *Self and community in a changing world* (2010), Kenyan philosopher D. A. Masolo discusses the communalist spirit and ethic among the Luo people of Kenya, citing Gabriel Omolo. Sometime in the 1960s, Omolo recorded an album with a Luo song titled *Jadak kende-ni m'unene en ng'a m'oyuma miwuoro*, which Masolo translates as "The isolated and autonomous individual is like a proverb, for he/she develops no sense of rules or obligations." (135) Such persons lead lives oblivious to their surroundings, as if they were the only ones who exist. While they may believe they protect their selves from others by cocooning themselves in their own worlds, their behavior is actually unfair and injurious, both to the community and to their own selves.

Masolo also engages with the Ghanaian philosopher Kwasi Wiredu, who emphasizes the importance of communication in human society in general (Masolo 2010, 134). Wiredu argues that communication is so vital to society that without it, the concepts of community and personhood become meaningless. As Wiredu notes,

> No human society or community is possible without communication, for a community is not just an aggregate of individuals existing as windowless monads but of individuals as interacting persons … Without communication there is not even a human person. (Wiredu 1996, 13)

The recurrent theme is that African societies are communalistic. It is this aspect that differentiates and distinguishes them from other so-called modern societies; it is this aspect that underlies their ethics, epistemology, logic, metaphysics, ontology, psychology, notions of punishment, and even language (Tempels 1969, van der Walt 1997). As J. Mbiti et al. point out,

> In place of Descartes's 'I think therefore I am,' we find in ... the black tradition, 'I am because we are; and since we are therefore I am. (Mbiti et al. 1995, 8)

It is the community that gives meaning to the individual. F. Hord and J. Lee illustrate this further:

> [The mainstay of communalism in traditional African culture] is the idea that the identity of the individual is never separable from the sociocultural environment, neither from the wider environment and nature – not even from the continuity of generations. Identity is not some Cartesian abstraction grounded in a solipsistic self-consciousness ... (Hord and Lee 1995, 7 – 8)

Ontologically, cosmologically, spiritually, and normatively, the individual is connected to the community, which is at the center of every activity, practice, belief, and value. Community, here, is not a mere conglomeration, but a tight composite of individuals. Ifeanyi Menkiti precisely expresses this in that 'we' is a 'thoroughly fused collective' (Menkiti 1984, 179). Polycarp Ikuenobe adds to this that the 'we' in African cultures is a transcendental or organic 'we' and cannot be reduced to its component parts. The individual self is, by various quasi-organic processes, constituted by the community, and the community is an organically fused collectivity of the individual selves. This 'we' is a prism of traditions, structures, interests, values, and beliefs that cannot be pinned down or reduced to the set of individuals and institutions constitutive of the community (Ikuenobe 2006, 56).

The Luo phrase '*ero kamano*' illustrates the inseparability of 'I' and 'we' in Africa cultures. The Luo would say *ero kamano* to someone who performed an act considered noble. In some ways, the phrase is an expression of gratitude. So one might think that *ero kamano* is Luo for 'thank you'. But the literal translation of the phrase is, 'that is how it ought to be.' The subtle semantic twist is understandable in the context of 'I am because we are; and since we are, therefore I am'. An individual is supposed to engage in acts consistent with the general wellbeing of persons. And when someone acts accordingly, the person will have done the right thing, and a Luo will say *ero kamano*. It is thus not so much an expression of gratitude but an acknowledgment of an action in accordance with communal expectations on reality.[8]

John Mbiti remarked that "Africans are notoriously religious" (Mbiti 1969, 1). Religion permeates all aspects of existence. When Africans revere a tree, participate in initiation rites, sacrifice animals, and pour libations to ancestors, they engage in a symbolic relationship with other persons, living and dead, or with 'others' in a more general sense, such as trees, because of their strength, longevity, and stability. There are no irreligious persons in traditional Africa. Living without religion means living outside the community, and amounts to somehow excommunicating oneself (Mbiti 1969, 2). Concomitantly, every object in the universe and all human activities have religious significance:

> Names of people have religious meanings in them; rocks and boulders are not just empty objects, but religious objects; the sound of the drum speaks a religious language; the eclipse of the sun or moon is not simply a silent phenomenon of nature, but one which speaks to the community that observes it, often warning of an impending catastrophe. (Mbiti 1969, 15)

Whatever European missionaries and scholars might have thought about the religiosity of Africans, there is a need to reevaluate the Western perception of African *weltanschauung*. This perception was informed by a belief in the superiority of the West, a belief that is now discredited by its perpetration of a planetary crisis. The African worldview, we believe, has a role to play in averting the threats and solving problems facing humanity today.[9]

So the traditional African notion of personhood, as explicated above, should be given more credence in our struggle to contain the adverse effects of environmental degradation and climate change. This notion brings to the table of world culture an ethics that is reciprocal, communal, and holistic, and it points to a vision of development that is neither linear nor individualistic.

Our explanation of communalism as a defining feature of African culture in general is not intended as a factual description of each and every African culture in particular. Individualistic elements have invaded Africa today to such an extent that it is difficult to say, on an empirical level, whether the cultures are (still) communalistic. Communalism remains strongest in the countryside, while urban centers are fairly cosmopolitan with explicit individualistic tendencies. But such modern changes should not distract us from the traditional African heritage, and from its potential, as has been argued here, of guiding the global adaptation to climate change. Climate ethics must include a communal dimension, and this dimension is clearly articulated in African sagacity.

## 6. African sagacity as climate ethics

Philosophic sagacity or sage philosophy originated in academic African philosophy in the 1970s. It was initiated and given visibility by the Kenyan philosopher Henry Odera Oruka. Initially, it was a response to the two dominant approaches in African thought, ethno-philosophy and professional philosophy (cf. Odera Oruka 1981, 1983; Ochieng'-Odhiambo 2010). Since then, it has been interpreted in different ways and even adjusted by some scholars to suit their objectives and aims despite the fact that they might, on the face of it, seem to deviate from the original function of sagacity as defined by Odera Oruka. For purposes of this chapter, we adopt Odera Oruka's original definition and interpretation. (For some of the various interpretations and shades of sagacity see Ochieng'-Odhiambo 2002 and 2006).

Sagacity consists of thoughts having insight and showing good judgment. It is the reasoning of persons acknowledged as sages by their communities. In another sense, sagacity is a body of principles or tenets that underlies the beliefs, customs, and practices of a culture. This second sense implies the first, since it is the reasoning of those acknowledged as wise that constitutes, in essence, the basis of communal culture (Odera Oruka 1991b, 33). In Odera Oruka's definition, sage philosophy and sagacity are synonymous:

> [Sage philosophy is the sum of] the expressed thoughts of wise men and women in any given community and is a way of thinking and explaining the world that fluctuates between popular wisdom (well-known communal maxims, aphorisms and general common sense truths) and didactic wisdom (an expounded wisdom and rational thoughts of some individuals within community). While popular wisdom is often conformist, didactic wisdom is at times critical of the communal set up and popular wisdom. (Odera Oruka 1991b, 33)

Sage philosophy has two components: popular or folk sagacity, and philosophic or didactic sagacity. The former comprises communal maxims, aphorisms, and common-sense truths; the latter consists of the expounded wisdom and rational thoughts of an individual in the community. Unlike his philosophic counterpart, the folk sage operates squarely within the confines of his culture, and the truth-claims within this culture are treated as absolutes that are not to be questioned. Anything outside of, or contradictory to the culture is treated with indifference, even hostility. Those sages who are merely experts in the culture defend the structure of their society with the zeal of fanatical ideologists (cf. Odera Oruka 1991a, 52).

Philosophic sagacity is a more general project to bring back some of the lost glory of philosophy. In his earlier essays, Odera Oruka defines a sage simply as a person "versed in the wisdoms and traditions of his people" (1983, 386). In his later work, he adds an ethical component to the definition, to underscore the practical aspect of philosophic sagacity. The reasoning of sages must be seen primarily as concerned with the moral and empirical issues relevant to the society. The sage is expected to offer insightful solutions to some of the issues. In this revised definition, a sage has two qualities: insight and ethical inspiration.

> So a sage is wise; he has insight, but employs this for the ethical betterment of the community. A philosopher may be a sage and vice versa. But many philosophers do lack the ethical commitment and inspiration found in the sage … A sage, proper, is usually the friend of truth and wisdom. A sage may suppress truth only because wisdom dictates not because of some instrumental gain. Indeed, Pythagoras' definition of a philosopher as the "lover of wisdom" should have been reserved for a sage, since the sophists were the grave-diggers of wisdom and truth. Socrates was wrongly labeled, "philosopher"; he was first and foremost a sage. Socrates used philosophy only as a means to advance his sagacity and expose the hypocrisies of his time. But when all is said, one must still emphasize that sagacity and philosophy are not incompatible. (Odera Oruka 1991b, 9-10)

While scientific and technological advances obviously benefit human lives, they also and equally obviously do a disservice to humanity in some cases. When that happens, science and technology become subject to ethical critique. This critique is not necessarily antagonistic. Genetic engineering, for instance, is far too important to be left solely in the hands of the scientific, engineering, and medical communities. Ethicists can perform vital services in guiding and regulating such advances. In recent times, however, a rift between the scientific community and the community of ethicists appears to have opened up, in that science and technology seem to have spawned their own kind of mercenary morality. This *technological morality* is an ethos in which technological innovations are intrinsically worthy of respect, and in which technological superiority or efficiency are identified with the good. Odera Oruka, therefore, rightly believed that if the thoughts of the sages were granted more intellectual and social spaces in modern Africa, then this would be one way of combating the detrimental impact of technological morality on African cultures; a morality that is adversely affecting all members of the global community as well.

In Africa today, the development of human wellbeing is increasingly identified with technological and economic development. On the individual level, this corresponds to an equation of personal improvement with the acquisition of the latest technological gadgets, fashions, and things. Development, which has also an organic and spiritual dimension, is increasingly reduced to material accumulation. This erroneous identification of the development of wellbeing with technological-economic development is evidently not only a problem for the rapidly modernizing African societies; it is a global issue. Examining this problem in Africa is guided by our interest in explicating the relevance of African philosophic sagacity. If well-articulated, properly substantiated, and readily available to community members especially in urban areas, sagacity could serve as a check on our self-centered and risky belief in the saving graces of technology.

With regard to the climate crisis, three points are worth noting about philosophic sagacity:

(1) What distinguishes a sage from a conventional academic philosopher are an insightful ethical quality or inspiration, and a concern for practical empirical issues. These are necessary to a sage, but contingent for a conventional philosopher.
(2) The sage also plays a vital role in defending the native culture from foreign elements that are ethically undesirable. Unlike conventional philosophers, sages are custodians of the cultural survival of their societies.
(3) Nonetheless, sagacity and conventional academic philosophy are not mutually exclusive. In fact, the ideal scenario would be to have both as complementary elements, jointly advancing a common cause, such that philosophy elucidates sagacity, and sagacity guides philosophy lest it becomes too abstract.

Considering this general function of the task of philosophic sagacity, it is apparent that it plays a role in our struggle with the adverse effects of climate change. On national-political and interdisciplinary-technical levels of intervention, the philosophic sage ought to be a participant, to ensure that other agents in the climate fight do not lose sight of holistic values and the cultural and existential bottom line. Some conventional academic philosophers are already engaged in this fight, discussing issues such as the compensation of climate change refuges, the applicability of the Maximin principle, global climate justice, the fair distribution of burden sharing, risk assessment and possible outcomes, and the ethics of cap-and-trade strategies. Such work is certainly useful, but it is also abstract, analytic, and legalistic. Philosophic sages contribute an earthier and more comprehensive perspective, not as far removed from practical day-to-day problems, not the least because their communities feel the effects of climate change firsthand. Sages complement the analyticity of conventional philosophical participants by representing a *synthetic* viewpoint, a perspective that integrates our social, natural, and wider environments. Ultimately, sagacity injects a dose of sanity in the arcane policy discussions, contributing a wider wisdom that goes beyond mere contracts, conventions, and technicalities of implementation.

Unfortunately, in African societies these days sages are not accorded the respect they deserve and once enjoyed. This is largely because today, in the name of modernity, too much value is attached to paper qualifications at the expense of the *ethical quality* of the professional. The more paper qualifications one has, the more one is held in esteem; the better chance one stands to be appointed to positions of influence and to high profile posts in international commissions. Since a good number of African sages do not have paper qualifications they often miss out on such appointments. Yet the truth is that, perhaps, they could do a better job than paper-qualified professionals. This contemporary marginalization of sages explains, in part, the rampant corruption and unethical practices that bedevil several countries in postcolonial Africa. We believe that this also plays a part in the climate policy quagmire. If failed conventions such as COP-15 Copenhagen and COP-17 Durban had a reasonable representation of philosophic sages – and the will to heed sagacious reasoning – then chances are there would have been more honesty and goodwill in the course of the deliberations, and ultimately considerably more success in the outcome. Instead of the sorry chaos of horse-trading, platitudes, and mutual recriminations witnessed by the participants, the meetings would have been guided by a real desire for mitigation and sustainability.

## 7. Conclusion

We have argued that the communal humanistic attitude that finds expression in the concept of personhood in African cultures should have a functional representation and articulation in climate ethics discourse. We also want to assert that the cardinal principles in African cultures – communalism and humanism –constitute an integral aspect of African philosophic sagacity. Another aspect is its rational dimension. In current climate ethics discourse, we have no doubt that the rational dimension, the philosophical side of sagacity, finds adequate expression, but we are concerned with the neglect of the sagacious side. Climate ethics, at this point, is perfectly reasonable, to be sure, but it lacks wisdom and the courage to ground thinking in community and humanity. That dimension still waits to be incorporated into the discourse. Plato remarked that "there will be no end to the troubles of … humanity until philosophers become kings or until … kings really and truly become philosophers" (Plato 2007, xxxii). Replace "philosophers" with 'climate ethics', and "kings" with 'sagacity', and then let the phrase be as follows: *there will be no end to the troubles of humanity until climate ethics becomes sagacious, or until sagacity really and truly becomes part of climate discourse.*

There is a fundamental need to change collective attitudes on what the quality of life ought to amount to, and what kind of development we are aiming for. The development paradigm based on the sanctity of continuous growth has turned out to be unsustainable. The invidious consequence of this paradigm we now experience as global warming and climate change. Soon we will reach the end of development, as understood in this linear model, unless we complete a paradigm shift geared towards a holistic view of the world, which includes duties towards our community and our environment as essential elements of well-being. Since there is no reciprocity of rights and duties with future generations, we must take our duties more seriously and cultivate a wider wisdom. One path would be to look more deeply into the original communal view of existence that comes out of Africa and its wisdom, which connects individuals, communities, tribes, and nature into one overarching functional entity that all are obliged to care for.

## Notes

1 The Bruntland Commission (1987) defined 'sustainable development' as 'development that meets the needs of the present without compromising the ability of future generations to meet their own needs'.

2 The World Commission on Environment and Development (WCED), for example, sees the "possibility for a new era of economic growth, one that must be based on policies that sustain and expand the environmental resource base. And we believe such growth to be absolutely essential to relieve ... poverty" (p.1). According to the WCED, better management of technology and social organization will ensure the desired growth (cf. p.8).

3 The United Nations Conference on Environment and Development (UNCED) in Rio de Janeiro, Brazil, 1992, known as the Earth Summit, resulted in the *Rio Declaration.* The declaration specifies sustainable development as the new goal of national and international policy. A key point at Rio was the recognition of the inseparability of development and environment; also, the highly developed countries in the North acknowledged their special responsibility for the environmental crisis. However, the declaration was issued without binding authority, and since then, no serious action has followed. Another diplomatic outcome of Rio was the *Framework Convention on Climate Change* (FCCC), the first treaty concerned with the rising atmospheric concentration of greenhouse gases (GHG) and its climatic consequences. Various emission caps and upper limits were debated over the years, most prominently in the first update to the FCCC, the Kyoto Protocol. The consensus that has emerged since Rio that the concentration of GHG measured in carbon dioxide equivalent units ($CO_{2equiv}$) needs to be lowered to 350 parts per million if dangerous climate change is to be avoided. The 2011 $CO_{2equiv}$ levels are 392 ppm. Negotiations from Rio to Copenhagen all ended in deadlock because the USA blocked binding targets for stabilizing and reducing GHG emissions. The 2012 follow-up conference to Rio, Rio+20, has reaffirmed the original goals but failed once more to make them binding. In *The Guardian* on June 22nd, 2012, G. Monbiot sums up the disappointment: "Several of the more outrageous deletions [to the 2012 Rio+20 text to be adopted] proposed by the United States – such as any mention of rights or equity or of common but differentiated responsibilities – have been rebuffed. In other respects the Obama government's purge has succeeded, striking out such concepts as 'unsustainable consumption and production patterns' and the proposed decoupling of economic growth from the use of natural resources".

4 On environmental ethics in Africa see for example Kelbessa 2005, and 2011, also Calestous, Mugabe, and Kameri-Mbote, 1995.

5 There have been several studies on the effects of climate change not only to the lives of the people in various African countries but on the overall environment. See for example Adger, Huq, Brown, Conway and Hulme 2003; Hulme, Doherty, Ngara, New, and Lister 2001; Ochieng and Kituyi, 2000.

6 UNDP's Human Development Index and also the UN Millenium Development Goals attempt to measure development in a wider sense that takes into account the multi-dimensional nature of poverty and quality of life better, but they still are based on the linear concept of development that is conditioned on continuous economic growth.

7 Léopold Senghor is renowned for his characterization of the subject-object dichotomy in the Western worldview: "He [the European] first distinguishes the object from himself. He keeps it at a distance. He freezes it out of time and, in a way, out of space. He fixes it, he kills it. With his precision instruments he dissects it in a pitiless factual analysis. As a scientist, yet at the same time prompted by practical considerations, the European makes use of the *Other* that he has killed in this way for his practical ends. He makes a *means* of it. With a centripetal movement he assimilates it. He destroys it by devouring it". (Senghor 1964, 72)

8 Placide Tempels makes a similar observation with regard to the Baluba, a Bantu group in the lower Congo of Central Africa. In emphasizing the cardinal role played by ontology among

the Baluba, he argues that besides ethics, their epistemology, psychology, religion, notion of punishment are all explainable and make sense only in terms of their ontology (Tempels 1969).

9 Upon attainment of political independence in the 1960s and '70s, leaders of several African nations attempted to revert to the cardinal principles of traditional Africa such as communalism. It was thought that reverting to these principles was the best way for Africa to free itself from the yoke of colonialism. Julius Nyerere, the founder of modern Tanzania, argues that in the communalistic spirit in pre-colonial Africa, the individual was rich or poor only to the extent that the society was rich or poor (Nyerere 1968, 17). Kwame Nkrumah, the founder of modern Ghana, calls for a social revolution, in which African thought turns to the redemption of the communalistic society of the past (Nkrumah 1970, 79). Despite their efforts to revive communalism in the emergent social political realities, their countries did not fare as well as they had hoped for. Their failure was, in part, due to interference by the West and, in Nkrumah's case, to betrayal by neo-colonialist reactionaries (Nkrumah 1970, vii). That their (Nyerere's and Nkrumah's) efforts languished should accordingly not be construed as an inherent weakness in communalism.

## References


Adger, W., S. Huq, K. Brown, D. Conway, and M. Hulme. 2003. Adopting to climate change in the developing world. *Progress in Development Studies* 3, no. 9: 179 – 195.

Appiah, K. 2006. Cosmopolitanism: Ethics in a world of strangers. New York: W. W. Norton.

Assadourian, E. 2012. The path to degrowth in overdeveloped countries in *Worldwatch. 2012. State of the world. Moving toward sustainable prosperity,* http://blogs.worldwatch.org/sustainableprosperity/wp-content/uploads/2012/04/SOW12_chap_2.pdf

Bentham, J. 1962. *Works*, ed. John Bowring, London, 1838 – 1843; reprinted New York: Russell & Russell, 11 vols, vol. 3, 159.

Calestous, J., J. Mugabe and P. Kameri-Mbote, eds. 1995. *Coming to life.* Biotechnology in African economic recovery. Nairobi: ACTS Press (African Centre for Technology Studies) Nairobi and London: Zed books.

Escobar, A. 1985. *Encountering development: The making and unmaking of the Third World.* Princeton: Princeton University Press.

Esteva, G. 1992. 'Development'. In *The development dictionary*, ed. W. Sachs. London: Zed Books.

Helliwell, J., R. Layard and J. Sachs. 2012. *The world happiness report.* The Earth Institute, Columbia University, CIFAR (Canadian Institute for Advanced Research, and Centre for Economic performance. http://issuu.com/earthinstitute/docs/world-happiness-report.

Hellsten, S. 2008. Failing states and ailing leadership in African politics in the era of globalization: Libertarian communitarianism and the Kenyan experience. *Journal of Global Ethics* 4, no. 2: 155 – 169.

Hellsten, S. 2011. What can African philosophy o er to cosmopolitanism as an ideology of freedom in the era of globalization? In *Conversations in philosophy: Knowledge and freedom,* edited by F. Ochieng'-Odhiambo, R. Burton and E. Brandon, Newcastle: Cambridge Scholars Publishing.

Hord, F. L., and J. S. Lee. 1995. *I am because we are: Readings in Black philosophy.* Amherst: University of Massachusetts Press.

Hulme, M., R. Doherty, T. Ngara, M. New and D. Lister. 2001. African climate change 1900 – 2100. *Climate Research* 17, no. 2: 145-168.

Ikuenobe, P. 2006. *Philosophical perspectives on communalism and morality in African traditions.* Lanham: Rowman and Little eld Publishers.

Janz, B. 2011. Watsuji Tetsuro, Fudo, and climate change. *Journal of Global Ethics (Special issue on Climate Change)* 7, no. 2: 173 – 184.

Kuo, Shih-yu. 2011. Climate change and the ecological intelligence of Confucius. *Journal of Global Ethics (Special issue on Climate Change)* 7, no. 2: 185 – 194.

Kelbessa, W. 2005. The rehabilitation of indigenous environmental ethics in Africa. *Diogenes* 50, no. 3: 17-34, http://dio.sagepub.com/content/52/3/17.

Kelbessa, W. 2011. Indigenous and modern environmental ethics: A study of the Indigenous Oramo environmental ethic and environmental ethics in the light of modern issues of environment and development. *Ethiopian philosophical studies I*, Washington, D.C. The Council for Research in Values and Philosophy.

Leduc, T. 2010. *Climate, culture, change: Inuit and Western dialogues with a warming North.* Ottawa: University of Ottawa Press.

Löfquist L. 2011. Climate change, justice and the right to development. *Journal of Global Ethics* 7, no. 3: 251 – 260.

Mbiti, J. 1969. *African religions and philosophy.* London: Heinemann.

Mbiti, J., E. Fahibusch, J. Pelikan, eds. 1995. *Internationale theologische Enzyklopaedie.* Goettingen: Vandenhoeck & Ruprecht.

Masolo, D. A. 2010. *Self and community in a changing world.* Bloomington: Indiana University Press.

Melkole, S. and A. Merriam. 1998. The Third World: De nitions and new perspectives on development. In *The new Third World*, ed. A. Gonzalez and J. Norwine. Boulder: Westview Press.

Menkiti, I. 1984. Person and community in African traditional thought. In *African philosophy: An introduction*, ed. Richard. A. Wright. New York: University Press of America.

Monbiot, G. 2012. Nowhere, nothing, nada: Rio+20 draft text is 283 paragraphs of u . *The Guardian*, 22 June.

Nkrumah, K. 1970. *Consciencism, philosophy and ideology for decolonization and development with particular reference to the African revolution.* London: Panaf Books.

Nussbaum, Martha. 1987. Nature, function, and capability: Aristotle on political distribution. *WIDER working paper 31*, Helsinki: UNU/WIDER.

Nustad, K. 2001 Development: the devil we know? *Third World Quarterly* 22, no. 4: 479 – 489.

Nyerere, J. 1968. *Ujamaa: Essays in socialism.* Cambridge: Cambridge University Press.

Ochieng, B. and Kituyi, E. 2000. Adapting to climate change: Challenges and options for Africa . *Innovation* 7, no. 2.

Ochieng'-Odhiambo, F. 2002. Evolution of sagacity: The three stages in Oruka's philosophy. *Philosophia Africana* 5, no. 1: 19 – 32.

Ochieng'-Odhiambo, F. 2006. The tripartite in philosophic sagacity. *Philosophia Africana* 9, no. 1: 17 –34.

Ochieng'-Odhiambo, F. 2007. Philosophic sagacity: A classical comprehension and relevance to postcolonial social spaces in Africa. *Quest* 21, nos. 1 – 2: 91 – 107.

Ochieng'-Odhiambo, F. 2010. *Trends and issues in African philosophy.* New York: Peter Lang.

Odera Oruka, H. 1981. Four trends in current African philosophy. In *Philosophy in the present situation of Africa*, ed. Alwin Diemer. Wiesbaden: Franz Steiner Verlag.

Odera Oruka, H. 1983. Sagacity in African philosophy. *The International Philosophical Quarterly* 23, no. 4: 383 – 393.

Odera Oruka, H. 1991a. Sagacity in African philosophy. In *African philosophy: The essential readings*, ed. Tsenay Serequeberhan. New York: Paragon House.

Odera Oruka, H. 1991b. *Sage philosophy: Indigenous thinkers and modern debate on African philosophy.* Nairobi: ACTS Press.

Plato. 2007. *The republic.* Trans. Desmond Lee. London: Penguin Books.

Pogge, T. 2002. *World poverty and human rights.* Cambridge: Polity Press.

Remenyi, J. 2004. Poverty and development. In *Key issues in development*, ed. D. Kingsbury, J. Remenyi, J. MacKay and J. Hunt. New York: Palgrave Macmillan.

Schussler, R. 2011. Climate justice: a question of historic responsibility? *Journal of Global Ethics* 7, no. 3: 261 – 278.

Schuurman, F. 2000. Paradigms lost, paradigms regained? Development studies in the twenty-first century. *Third World Quarterly* 21, no. 1: 7 – 20.

Sen, Amartya. 1984. *Poverty and famines.* Oxford. Clarendon Press.

Senghor, Léopold. 1964. *On African socialism.* London: Pall Mall Press.

Subramaniam, B., J. Bever and P. Schultz. 2002. Global circulations. Nature, culture and the possibility of sustainable development. In Feminist post-development thought. Rethinking modernity, post-colonialism and representation, ed. K. Saunders. London and New York: Zed Books.

Tempels, P. 1969. *Bantu philosophy.* Paris: Présence Africaine.

UNDP. Human Development Report 2011. Sustainability and equity: A better future for all. http://hdr.undp.org/en/reports/global/hdr2011/download/en/

United Nations Statistic Division. Millennium indicators. Monitoring progress towards the achievement of millennium development goals. http://milleniumindicators.un.org/unsd/mi/mi_highlights.asp (2004-2015)

United Nations Secretary-General's High-Level Panel on Global Sustainability (2012). Resilient people, resilient planet: A future worth choosing, Overview. New York: United Nations. http://www.un.org/gsp/sites/default/files/attachments/GSP_Report_web_final.pdf

Van der Walt, B. 1997. *Afrocentric or eurocentric? Our task in a multicultural South Africa.* Potchefstroom: Potchefstroom University for Christian Higher Education.

Verhoef, Heidi and Claudine Michel. 1997. Studying morality within the African context: A model of moral analysis and construction. *Journal of Moral Education* 26, no. 4: 389 – 407.

Walker, J. 2000. *Environmental ethics.* London: Hodder and Stoughton.

Wiredu, K. 1996. *Cultural universals and particulars.* Bloomington: Indiana University Press.

World Bank. Partnership in development. Progress in the fight against poverty. The World Bank Group, Washington, D.C., 2004. (2004-b).

World Bank (2011) Global monitoring report 2011. Improving the odds in of achieving the MDGs. http://siteresources.worldbank.org/INTGLOMONREP2011/Resources/7856131-1302708588094/GMR2011-CompleteReport.pdf

World Commission on Environment and Development. 1987. *From one Earth to One world: An overview.* Oxford: Oxford University Press.

Worldwatch. 2012. State of the world. Moving toward sustainable prosperity http://blogs.worldwatch.org/sustainableprosperity/sow2012/

Xia, Chen and M. Schönfeld. 2011. A Daoist response to climate change. *Journal of Global Ethics (Special issue on Climate Change)* 7, no. 2: 195 – 204.

# Ancestral climate wisdom: Return to a thoughtful etiquette

Timothy B. Leduc

*Faculty of Environmental Studies, York University*

## 1. Introduction

'Why don't researchers ever ask us about wisdom?' Almost a year after I began talking with Jaypeetee Arnakak on Inuit ways of thinking about northern warming, he asked me this question. From his position as an Inuit policy worker and philosopher, Jaypeetee stressed that wisdom, or *silatuniq* in Inuktitut, should be of central importance to anyone concerned with climate change (Leduc 2010a). Considering the significant changes that are occurring globally and in the North, a region that some describe as climate change's 'canary in the mine', it may seem highly impractical to shift our attention from practical questions of how to reduce our greenhouse gas emissions to that of wisdom. What may seem even more impractical is the argument I am going to make in this chapter: that a sustainable and just response to northern warming and global climate change may depend on our capacity to inspire climate research and politics with a wise way of thinking that is at root an etiquette toward the surrounding world.

For many who study northern warming and global climate change, there is an increasing sense of urgency that a comprehensive response needs to be initiated now; the time for delays is over. Representative of this concern is Martin Schönfeld's recent global ethics analysis that led him to describe climate change as an approaching "fork in the road" of extreme consequence. He puts the choice this way: "Either civilization keeps working with the dominant paradigm and enter failure mode, or humankind learns how to change" rapidly and dramatically (2011, 131). While making social changes in the face of uncertain climate forces is, as we will see, a historically recurring story for humanity, Schönfeld also makes the vital point that a significant difference is that today's change is happening collectively on a global rather than localized scale. In my book *Climate, culture, change*, I likewise discuss this unique situation we find ourselves in through considering the recurring stories of apocalypse across cultures and times, including a present moment that sees them proliferating in Christian, environmentalist and economic guises. As with Schönfeld, I argue that apocalyptic stories ultimately present us with a choice. In my words, "Apocalypse can be a global cultural end that occurs at some specific moment in time, but on a more practical everyday level it can inform an adaptive ethical practice for facing a future that is continually 'breaking in and exploding every complacency'" (2010a, 151). It is exactly because we require an urgent *cultural* change in our etiquette, not simply adaptive management, that I argue our climate thought needs to be informed by something akin to Jaypeetee's *silatuniq*.

There is an interesting paper written by Jim Cheney and Anthony Weston on an ethics-based epistemology that illustrates some of the core issues I want to contemplate here in relation to the wise transformation of Western climate thought. They propose the interconnected nature of environmental issues like climate change radically call "into question basic assumptions concerning the relationship between epistemology (how we come to knowledge) and ethics and, hence, basic assumptions concerning ethics itself" (1999, 115). One of those dominant beliefs in need of questioning is the common Western view of ethics as a systematized set of responses to a world that is made "of a collection of more or less given facts" (1999, 116). It is our increasing recognition of the limits to knowledge in the face of a complex climate response to human activities that challenges such a fact-based ethic. Considering this still prevalent assumption, Cheney and Weston go on to ask some poignant questions: "What if the actual necessary relation is the other way around? *What if the world we inhabit arises most fundamentally out of our ethical practice, rather than vice versa*?" (1999, 115-6) Their basic premise is that environmental issues move us "away from epistemology-based ethics and toward an ethics-based epistemology" (1999, 123). In fact, it is more an environmental etiquette than ethic, for it contextualizes our knowledgeable thought within a mannered and respectful engagement of that which remains uncertain. What I similarly learnt through dialogue with Jaypeetee and other Inuit is that *silatuniq* is concerned with a way of living that is aware of the limits to knowledge in a deeply interconnected existence. Wisdom is a kind of etiquette for complementing and contextualizing what is knowable.

Though my dialogues with Inuit on *silatuniq* and other common research concerns like traditional ecological knowledge highlighted narrowing interdisciplinary trends and the value of broadening climate research, in the end I did not explicitly consider what interdisciplinarity means as a way of coming to knowledge in a wise manner and not just integrating knowledges. Perhaps there is some potential in the sociality of interdisciplinary and intercultural climate research that situates it at that crossroads where etiquette can begin to inform fact. This chapter is concerned with the historic re-emergence of a climatic uncertainty that *has* in the past and *is* in the present influencing new ways of thinking and living. It is an interdisciplinary inquiry informed by indigenous understandings of climate change like that of Inuit, research on the changing history of human-climate relations, and philosophical analyses like that of Cheney and Weston on the capacity of environment and climate to influence human thinking and ethics. As I will argue, the patterns that connect these different discourses have a lot to suggest about the manners of sustainable climate thought and current dominant tendencies in climate research and politics that are largely resistant to such drastic change. Our ethical fork in the road, I will argue here, calls for a return to a wise climate etiquette, one that can allow us to relationally think with our interpenetrating global and local realities.

## 2. Climate thinking

Researchers have been heading north to document Inuit observations of changing weather, land, ice and animals for decades now, and as with many I was originally focused on Inuit ecological knowledge of climate change before Jaypeetee shifted our discussion from this well-worn path to a consideration of *silatuniq*. For weeks, we discussed the relevance of *sila* to northern warming which he described as an ever-moving and imminent force that surrounds and permeates Inuit life, with it most often experienced in the weather. Contrasting his Inuit sense of *sila*, I brought to our dialogues knowledge from two largely divided academic disciplines. Contemporary climate research with Inuit was defining *sila* as a direct translation for weather, with it most often coming up in relation to unexpected weather phenomena. Meanwhile, Inuit ethnographies from the first half of the twentieth century described *sila* as the spirit of the air, upholder of the weather, and the breath source of all life. It was while discussing this divide in Western thought between the physical weather properties of *sila* and a more spirited sensibility that Jaypeetee brought up *silatuniq*. As he explained, the *sila* and climate surround our lives, and *silatuniq* is an inquiry into "the context and consequence of applying knowledge and/or how our interacting with the surround affects that surround". This understanding seemed relevant to northern warming and global climate changes that are at root a planetary response to industrial society's exhalation of greenhouse gases.

What I do not discuss in *Climate, Culture, Change* is that such Inuit concepts as *sila* and *silatuniq* began to motivate another set of questions on the methodological nature of inter/trans-disciplinarity in academic environmental thinking and studies. There seemed to be a resonance between Inuit *silatuniq* and the late-1960s rise in environmental interest that, in the academy, led to a host of writings on the idea that the surrounding ecology can shape and influence the nature of our thought. Though this idea stretches back to Aldo Leopold's (1970[1949]) mid twentieth century reflections in 'Thinking like a mountain' where he describes his initiation into a broader ecological thought than that of limited human motives, the first to coin the term "ecological thinking" was environmental philosopher Paul Shepard who wrote a seminal 1967 essay describing the self extended into "the landscape and ecosystem" (1967, 2). Not long after Shepard, there was also Gregory Bateson's influential *Steps to an Ecology of Mind* that similarly proposed "the mind is not limited by the skin" (1972, 454). Of particular relevance to my interest in *silatuniq* is the unfolding thought of Shepard which concluded that the long Palaeolithic history of hunter-gathering led the human mind to pattern itself after the animals, ecology and climate it depended upon. Seemingly intimating Inuit *sila* and *silatuniq*, he would write that ecologically-grounded indigenous approaches of past and present can "carry us beyond ourselves, pursuing the nature of thought as the thought of nature" (1996, 63).

As influential as these writers were on the idea of "ecological thinking" generally and my own thought as I engaged Inuit, my initial sense of such mind-ecology relations ultimately arose from an early pedagogical document in Canada's oldest Environmental Studies program at York University where I began graduate work in 1999. Founded in 1968, the program described its studies in the following way (Carrothers et al., 1968/1988, Foray Four 9):

The concept of 'environment' may be seen not only to mean the phenomena which are studied, but also to mean an inherent way of studying – environmental thinking.

The notion, relatively undeveloped [in the West], is that subject matter has an intrinsic form of thought as well as content, and that the environmental context of substantive concerns may be used to determine form of thought.

Influencing this description was John Livingston who similarly defined environment as the "context" we live and think within, and, his own specific field, "wildlife conservation not necessarily as an activity, but as a state of mind" (2007[1981], 12). One implication of this view is that inter/trans-disciplinary environmental research and pedagogy is seen as being responsive to the surrounding world and not merely a human innovation. In other words, disciplinary concepts, cultural understandings and institutional traditions – like those that limited my sense of *sila* – are decentred within a relational environment and climate that must inform our ways of thinking and living.

From very early on, there seemed to be a connection between such environmental or ecological thinking and the older Inuit sense of a contextualizing *silatuniq* that arises through intimate relations with the *sila*. While *silatuniq* is concerned with, in Arnakak's words, the "consequence of applying knowledge" in a deeply interconnected reality, the still "relatively undeveloped" Western idea of ecological thinking arose within a social context of increasing concern about human impacts on local and global environments. The story related by Leopold of watching the green fire seep out of the eyes of a wolf he killed while participating in an early twentieth century idea of resource management that attempted to eradicate the wolf is merely one example of how ecological wisdom has arisen from the experience of misapplied knowledge. In retrospect, he concluded that our incapacity "to think like a mountain" means the mountain lives in fear of not only deer unrestrained by the wolf's howl, but also those deeply intertwined human plans that are often grounded in short-sighted thought (1970[1949], 140). Turning to Bateson, he similarly writes that we forget the "eco-mental system called Lake Erie is a part of your wider eco-mental system – and that if Lake Erie is driven insane, its insanity is incorporated in the larger system of our thought and experience" (1972, 484). As with Leopold, he concludes the organism which cannot reflect the deeper and broader ecological thought that surrounds them is doomed to destroy that same environment which is the ground of their existence. Such unwise human thoughts as a world without wolves, lakes as pollution sinks and/or carbon as a tradable commodity has, in this way of thinking, unforeseen consequences.

It should be clear that ecological or climate thinking is not actually thought in the way it is dominantly conceived today as a largely internal mental reasoning, but rather some kind of relational epistemology and ontology that gives the world space to transform our ways of thinking as they are embodied in our actions – from the personal to institutional to political to cultural. I think it is in a context of this understanding that Cheney and Weston's (1999) differentiation of some main assumptions in the current dominant epistemology-based ethics and their alternative environmental etiquette is insightful. The first two assumptions in our dominant approach basically states the world is knowable enough to determine proper ethical action and that ethics is predominantly a response to what we know of the world. But from the perspective of ecological thinking, the deeply interconnected nature of human, ecological and climatic processes always severely limit our knowledge and thus our ethical relation to the world. We cannot "think like a mountain" because our knowledge is limited by various cultural assumptions concerning how we value a wolf. In relation to my climate research, it is the uncertainty surrounding climate change – from the way cultural predispositions demarcate climate research to the influence of climate denial politics on our responses – that highlights the continued relevance of Inuit *silatuniq* not just for the north, but also the practice of interdisciplinary climate research and global politics.

We need wise climate research and politics that is reflexively aware of "how our interacting with the surround affects that surround". This is where Cheney and Weston's alternative etiquette has some interesting points for considering a wise Western approach to climate thinking. Rather than ethics arising in response to a knowable world, the world itself is seen as filled with hidden possibilities that modify our ethical stance from knowing and managing to engaging and enriching the world. Critiquing the scepticism that afflicts both postmodern and analytic philosophical approaches to climate ethics, Schönfeld similarly proposes we need to move beyond the postmodern emphasis on culture over nature and the analytic focus on de-contextualized information. He writes both these ethical ways of thinking "appears wrong-headed and obsolete, because climate change asks philosophers to do the precise opposite: the reconciliation of information, the understanding of data, and the integration of events in context. Analysis is not needed now; synthesis is" (2011, 134-5). It is a view not inconsistent with Bateson's definition of ecological thinking as grounded in a complementarity methodology that is in "the nature of 'service' rather than dominance" (1972, 335). Lack of knowledge does not lead to paralysis of thought and action. Rather, it calls us to deeply engage the ecological and climate realities that surround us with an etiquette that can continually transform not only what we know but how we think and act in the world.

This emerging sense of climate thinking as primarily an etiquette for fostering complementary understandings can be seen as informing the relatively recent rise of both interdisciplinary and intercultural approaches to climate and environmental research. My dialogues with Inuit on concepts like *sila* and *silatuniq* required me to engage a wide range of discourses that included, but were not limited to, climate research, ethnographies, adaptive management, environmental ethics and religious studies. They also simultaneously highlighted the way in which contemporary climate research, politics and media are often constrained by a focus on the physical reality of the issue and economic viability of responding (see Leduc 2010a). This concern with interdisciplinary narrowness led me to my current project that is concerned with the scope of interdisciplinary research and pedagogy in the evolution of academic environmental studies and, as I have mentioned, the etiquette of ecological thinking. My first paper in this research critically analyzed some dominant program trends in academia and potential impacts on the complementarity of knowledge that influences ecological thinking (Leduc 2010b). Amidst five commentaries written in response to the article, Paul Woods suggested truly radical interdisciplinary environmental studies that can consider its own cultural limitations would need to be critically informed by the local indigenous understandings of where programs are situated. In his words, the future of environmental studies "is to educate students who are capable of imbibing some of the wisdom of indigenous cultures and who could, in time, offer effective counterweights to the hegemony of science" (2010, 46). It is not that indigenous understandings of ecology and climate are factually correct or impervious to limiting cultural, political and economic biases, but rather they have the potential to offer stark epistemological and ontological contrasts that can help us consider more adaptive ways of thinking and acting than what currently drives our lives.

## 3. An ancestral etiquette

There are indications in research that the knowledge sharing intent of today's interdisciplinary and intercultural climate research is a social adaptation with an old human ancestry. Since Shepard first connected ecological thinking to the influence of hunter-gatherer ways of living on humanity, other researchers have further developed our sense of how changing climates influenced the evolution of our hunter-gatherer minds and cultures. In the book *A Brain for all seasons*, W. Calvin (2002) synthesizes the latest research in paleoclimatology, human evolution and cognitive biology to offer an interesting view on the cycling impacts of extreme climate changes. While noting the way in which many of humanity's evolutionary jumps coincided with major cyclical climate shifts toward and away from recurring Ice Ages, he also clarifies it was the unsettled extreme periods of drought and flood rather than gradual temperature changes that brought the most significant stressors and challenges. In Calvin's words, there are signs the evolution of our minds, tool-making and social culture are primarily grounded in "the decade-long transitions between states, not the steady-states (warm-wet and cool-dry) themselves" because the former created enough disruption to manifest "relevant niche opportunities" (2002, 184). Extremes not only created more opportunities for marginal adaptations to come into play, their recurring nature ratcheted up our adaptive trajectories to produce "major changes in body and behavior in only a million years or so" – a process that Calvin defines as "catastrophic gradualism" (2002, 18). Such a pre-historic perspective resonates with Cheney and Weston's (1999) third critique of our dominant incremental approach to ethics which is based on the thought that change is slow and on the margins. The discontinuous nature of change reflected in climate and ecological research like Calvin's reveals to Cheney and Weston the need for an ethic that is socially mannered in the face of our limited knowledge and a world of uncertain dimensions.

The etiquette proposed by Cheney and Weston has another consistency with the research of Calvin that is directly relevant to our emerging sense of climate thinking. Considering the evolution of altruistic behavior and the common sharing protocols found across hunter-gatherer cultures, Calvin writes the "repeated fragmentation of large prehuman populations into many smaller subpopulations" during periods of extreme change could occasionally create "a group with a critical mass of sharers" (2002, 93; also see Burroughs 2005). While the more individualistic and self-interested groupings wasted "time and effort at fighting over the remaining food, the groups that shared (and otherwise minimized conflict) might have survived better" (2002, 93). Similarly drawing upon the relation of major climate changes to human evolution, anthropologist Brian Hayden (2003) noted a pattern in research that suggests an emotion-based social, perhaps religious, etiquette evolving from this climatic context. He describes rituals in early hunter-gatherer cultures as an emotive connecting activity that, "create lasting bonds between individuals and groups", bonds that nurture alliances between groups and "help them survive periodic disasters" (2003, 31-32). This conjecture is based on a combination of the archaeological record and what ethnographers have commonly documented in twentieth century hunter-gatherer communities. It is just such ethnographic insights of an earlier period that influenced not only Shepard's indigenous approach to ecological thinking, but also Cheney and Weston's conception of their etiquette as being historically grounded in an older sensibility that equates knowledge with "intimacy and reciprocity (rather than fact)" (1999, 123-4).

There are interesting connections between this view on an evolving climate etiquette and the Inuit hunting-based *silatuniq* that Jaypeetee had me consider in relation to climatic animal changes and their influential ancestor known as *Sedna* – the one down below the water who is the Mother of all the animals (see Leduc 2010a). Rather than the environment being primarily scarce in the sense of being in need of economic maximization for optimal profit, *Sedna* is in this animated view seen as an abundant sharing cosmos wherein "as human agents appropriate their shares they secure further sharing" (2010a, 190). The animals that sustain Inuit arise from Her being and when there are animal disturbances, like those occurring today in the North, *Sedna* would be seen as responding to unmannered human behavior. Such a perspective led to an etiquette that seemed to inform the Inuit knowledge system, taboos and rituals, and the concept of *ningiqtuq* as documented by ethnographer George Wenzel (1995). This term refers "to a web of social mechanisms for ensuring food and other resources are distributed from the individual to extended families and the broader community" through hunting protocols, communal meals and other practices (2010a, 190). The sharing also has spiritual dimensions that stretch out beyond the human community, such as in the Inuit hunting ritual of giving a dead seal its last drink of water and shamanic séances to *Sedna* during times of animal disappearances. Rituals like these highlight the point that mannered climate thought is not merely focused on human relations, but is deeply connected to ever-cycling ecological and cosmological surrounds.

It should be pointed out that Inuit *Qaujimatuqangit* or knowledge of *sila* and *Sedna's* animal changes has been of interest to climate research because it contextualizes current ecological changes within a sense of what happened in the past – including hunter experience, elder memory, and ancestral stories passed through generations like those of *Sedna*. Inuit describe their knowledge as being "passed down to us by our ancestors, things that we have always known, things crucial to our survival, patience and resourcefulness" (2010a, 4). Just as ecological or climate thinking attempts to shape itself after a world it cannot fully know, Inuit knowledge attempts to reproduce *Sedna's* giving cosmos by reminding humans they "are part of a transcendent universe in which everything emanates from the same spiritual source" (Nuttall cited in 2010a, 191). This ancient *silatuniq* suggests we are deeply intertwined with not simply what we know about the world, but more fundamentally with the social etiquette that informs our life relations and climate thinking.

The intriguing idea that a mannered climate thinking is originally a religious evolution is stretched further into the past by Robert Bellah (2011) in his comprehensive and interdisciplinary *Religion in human evolution*. He documents evidence of social rituals evolving from empathic behaviors that "began much simpler, with the synchronization of bodies, running when others run, laughing when others laugh, crying when others cry, or yawning when others yawn" (2011, 72). This 'motor mimicry' fostered social bonds and gradually became increasingly complex as people came to further ritualize their interconnection to surrounding communal, ecological and cosmological realities. In fact, Bellah proposes that ritual evolved initially as an embodied musilanguage or 'vocal grooming' that could allow bonding of "larger groups than was possible using the conventional primate mechanism of physical grooming" (2011, 123; also see Rappaport 1999). Returning to Hayden, he holds that the powerful emotions arising from ritually "communicating with common deities or totemic ancestors" further strengthened communal bonds, the capacity of communities to survive and the "penchant for religious experiences" across human cultures to this day (2003, 32). While such ritual communing is, in Bellah's words, "indeed in our heads, it is still very much in our bodies" (2011, 72). In other words, climate thinking emerges from an etiquette that socially relates us to the beings we bodily, culturally and religiously live with. It is a view consistent with Bateson's words that our mind is immanent in the "pathways and messages outside the body", and his religious speculation that this "is comparable to God and is perhaps what some people mean by 'God'" (1972, 461). Fostering an empathic relation with a sharing cosmos that has both knowable and unknowable dimensions may have been the source of our original etiquette for climate thinking.

Rituals and stories that intimate social relations with the climate are still with us today, and not only in indigenous traditions like that reflected in *Sedna*. There are a host of myths found across a wide range of cultures that have climatic dimensions and which talk of uncertainty and social choices. There has, for example, been a wide interest on the way in which our most recent planetary transition from an Ice Age may be behind an array of flood stories told from the Mediterranean to North America (e.g. Issar and Zohar 2004; Wilson 2001; Deloria 1994; Thompson 1981). As Wilson explains, human populations have for a long time settled on flat land along seashores and when, following the last Ice Age, "breakthroughs of the sea occurred, quite possibly suddenly and unexpectedly, they were accompanied by serious localized catastrophes" represented in flood stories told worldwide (2001, 8 – 9). From an indigenous perspective, Lakota scholar Vine Deloria states the global prevalence of flood experiences "gave rise to the legends that recorded memories of the flood" as well as encoding "moral and ethical codes particular to the religion reporting the flood experience" (1994, 72). While such myths tell of real uncertain events that deeply marked people's lives, they have, as Deloria points out, ethical implications for those who engage them because they associate these changes to cosmologically improper human behavior. Looking at the flood story told by Utnapishtim in the *Epic of Gilgamesh*, Thompson finds a message that says we cannot save ourselves "from the catastrophes that always come at the end of a world cycle, one cannot simply escape", but rather must be attuned to the emerging messages of the surrounding ecology and cosmology (1981, 202). It is interesting to note that, as with Calvin's "catastrophic gradualism", Thompson views such myths as arising from cycling climate transitions between stability and extreme uncertainty.

Myths like those of the flood reflect the profound effect that extreme climate changes had on human ways of thinking with the world, and as such can – if contemplated in more depth than this brief survey of flood myths allows here – offer profound insights on how to make our way toward sustainable thought and ways of living today. Some myths I have spent more time with in my book include Inuit stories of *Sedna* and Western views of apocalypse. In a time of climate change, one can hear intimations of apocalypse in stories that range from the fundamentalist Christian vision of hellish brimfire for the unbelievers to secular environmental concerns of societal collapse in the wake of interacting climate change and peak oil scenarios (see Leduc 2010a). While in the Christian tradition a universal or global 'End' is a common feature of apocalypse, indigenous North American stories tell of a moral cosmos that is responsive to human behaviors and whose cycling from destruction to renewal mirrors "the cycles of the seasons with its endless circle of degeneration, death, and rebirth" (2010a, 153). What these culturally and temporally ranging apocalyptic stories have in common is that they "are cultural strategies for understanding and responding to the uncertainties in people's lives" (2010a, 153), and as we have seen there has been a recurring climatic cycle of such uncertainties that conform closely to the indigenous sense of endless cycles.

The current prevalence of apocalyptic stories in various cultural guises is by no means a new phenomenon either, for our current time of change harkens back to the cycling "experience of disruptions whose uncertainty" led people to story them in apocalyptic myths that arose from physically powerful and spiritually numinous experiences (2010a, 153). It is a concern with just such recurring uncertainty stretching into the deep past that informs Calvin's portrayal of our climatic challenge: "Use our civilization's achievements to prevent the next whiplash, or lose much of civilization's gains as our warm period suddenly ends with a population crash" (2002, 295). The cycling from Ice Age to stable warm periods like the Holocene and back to Ice Age leads him to be concerned not simply with the role of human greenhouse gas emissions on a warming trend, but the capacity of those emissions to de-stabilize the stable warm pattern of our waning Holocene period and thus bring us into a period of rising uncertainty not unlike those experienced by our ancestors many times. Though we must reduce our emissions, Calvin also proposes our goal "must be to stabilize the climate in its favourable mode and ensure that enough equatorial heat continues to flow into the waters around Greenland and Norway" (2002, 291). This, he argues, can be designed for "in computer models of climate, just as architects design earthquake-resistant sky scrapers", adding that "we may not have centuries for acquiring wisdom, and it would be wise to compress our learning into the years immediately ahead" (2002, 291). We may be entering a period of transition from the climatically stable Holocene to some new epoch on the other side of industrial warming's instability, and our paradoxical challenge is to transition into some type of dynamically conceived steady-state culture.

Stories of a moral cosmos like those of *Sedna* and apocalypse lead me to take a different *silatuniq* out of Calvin's case study than an intensification of rational efforts at managing the planet. It is interesting to note that what led Livingston to define "wild-life conservation as a state of mind" was a recognition of the way in which conservation and environmentalism was failing due to its internalizing of an economizing mindset and management ethic that still primarily defined ecology "as a human asset" (2007[1981], 166; 1994). He was not alone at the dawn of the environmental movement raising such concerns, for Bateson similarly concluded it was "doubtful whether a species having both an advanced technology and this strange way of looking at its world can endure" (1972, 337). We are caught in a pervasive worldview and way of living that Bellah even associates with a reductionistic sense of our evolutionary history. As he explains, we have "become fixated on growing technological sophistication as the key to understanding human evolution" because it fits "our penchant toward economic determinism in the understanding of history", but it is becoming increasingly clear "that technology is more an indication of increasing cognitive capacity than a primary determinant in its own right" (2011, 126). In other words, the multidimensional social relations that have shaped our minds is that which opens up space for technological innovation, rational science and economic management. This inverted vision deeply contrasts our current dominant predisposition of holding technology up as the eventual panacea of all our climate and environmental crises, despite these solutions being fundamentally grounded in a still globalizing self-interested ethic that is consuming our future as fast as our technology allows us to.

Though there is obvious resonance between paleoclimate research like that of Calvin and the persistence of apocalyptic stories, it remains difficult for Western researchers, myself included, to truly embrace the latter's wise sense of a moral cosmos whose apocalyptic potential requires our climate thought and politics to be primarily socially mannered. It is easy in today's highly anthropocentric culture to assume that the evolution of communal sharing was simply a rational response to climatic uncertainties, but I think such a sensibility misses the fundamental point that we are grounded in a responsive and cycling cosmos. For Inuit, *Sedna* is an 'ancestor' connected to their own existence. In Bellah's research, the earliest religious impulse was toward an ecological, climatic and cosmological surround filled with ancestral and divine beings. He writes that this idea of ubeity – the interpenetration of physicality and spirit – "is difficult for us to grasp, yet it may be central to the way of life of tribal and archaic peoples" (2011, 158). Even the scientific story of evolution tells of our deep ancestral connection to all species on this planet and in the cosmos as epitomized in scientific views of biophilia (Wilson 1984) and Gaia (see Leduc 2010a; Primavesi 2009). Sharing is not simply an innovative human ritual. All of this recalls for me Thompson's description of myth as "a reworking of what was ancient for the ancients," that being a holistic sense of "the beginning and ending of things" which so contrasts both the progressive linear power of scientific reductionism and never-ending economic growth (1981, 210). Knowledge is most definitely needed for navigating a sustainable way forward in the midst of today's arising uncertainties, but if our rational capacities are not contextualized in a mannered *silatuniq* then our thinking and politics in a time of climate change will inevitably be unsustainable.

## 4. Conclusion: Pre-cautionary wisdom

> Swaggering, talking too loud, not knowing how to listen, this very clumsiness we now re-conceive as the fundamental ethical failure: failure to acknowledge and understand ourselves as living in a larger animate universe, and failure to co-participate with that very universe. Instead, we drive it into silence, and then take that silence to confirm our own centrality. - Cheney & Weston 1999, p.134

Perhaps the common reverence for ancestors documented across cultures reflects in a variety of ways ancient human insights about the value of a mannered approach to a social cosmos, especially when the latter acts with great uncertainty. The world is filled with messages that can inspire our thought beyond what we rationally know, but I have argued our current predispositions are limiting our capacity to embody an etiquette for being so engaged. Reflecting on Bateson's thought, Thompson (2001) concluded we are talking primarily about ontological and secondarily epistemological discrepancies between conscious societal purpose and larger ecological pathways that continually elude rationalization. While I have approached this issue by re-conceiving apocalypse as "an adaptive ethical practice for facing" a recurring climatic uncertainty (see 2010a, 169), a comparable religious sense of our ontological challenge is offered by Thompson. He writes it may be impossible for our thinking to ever "become identical with being, or perhaps it can only become so with the achievement of Buddhist Enlightenment" (2001, 17 – 18). To participate in a socially dynamic reality that always transcends our knowledge does not mean our search for knowledge and sustainable ways of living ends, just that those enlightened searches need to be conducted with an active recognition of the animate mystery that permeates all our ways of living and thinking. Not simply because we recognize the limits to our knowledge should we be cautious, but because the world is socially animated and responsive we need to be mannered in our approach – our pre-caution comes from that respectful engagement of the world. This is the mythic dimension of not simply a precautionary ethic, but a wise pre-cautionary etiquette for living and thinking with a climatically changing cosmos.

When I look at the political climate I live with today in Canada, it seems as though such climate manners is at best antithetical and at worst sacrilegious to the swaggering self-interested ethics of our politicians. Representing us on the global stage, Prime Minister Harper has effectively moved the national debate from the nature of a climate change response to the economic necessity of building pipelines from the Alberta tar sands to the highest bidder – whether that be the United States or China. Closer to my home on the municipal scale of Toronto, our newly elected conservative mayor Rob Ford has declared "the war on the car is over" and in the process has brought great uncertainty to the city's future mass transit plans and climate response. During a time of economic uncertainty, a large proportion of the population, though not a majority, have elected leaders who act unilaterally and are primarily concerned with driving the Canadian and international economy toward a model of growth based on more unfettered individual self-interest. As I discuss in my book, such anti-social political tendencies are deeply rooted in Canada's colonial history as a frontier economy that began with the fisheries and fur trade and continues today in the form of the tar sands (see Leduc 2010a). Confronting this national mindset and way of living that has deep North American, European and international dimensions is increasing evidence that individual self-interest was a failure during past periods of climate instability, and it is highly likely that our current global approach to trickle down self-interest will ecologically and climatically fail our common future.

We are fixated on an evolutionary backwater that cannot provide us with a sustainable way of thinking or living, and it increasingly seems like we are going to go through a lot of pain before truly re-considering our choice of direction at this significant fork in the road. But that is only the pessimistic side of the story, for we have seen there is also a more ancient etiquette that we can attend in our search for an alternative path in the midst of our climatic and social instability. That this alternative is not out there in some other culture, but in our common human ancestry, is a vitally important point. For if we follow the thought of Hayden, evolutionary transformations do not arise as miraculous mutations but rather "work most effectively where there is already some incipient expression of a trait that can become adaptive under changed conditions" (2003, 25). Perhaps we can see the current rise of interdisciplinary, intercultural and community dialogues occurring on global and local scales in response to our increasingly uncertain times as just such an "incipient trait," one that could become more adaptive as our self-interested maladaptations continue to crack in the face of economic, social, ecological, climatic and cosmological forces.

Beyond such social gathering of knowledge, we also need to engage a *silatuniq* practice that can socially relate one's mind with the mythic dimensions of the mystery that surrounds us. While Thompson describes a way of paraduction that lets "the mind wander to gather what associations it may, through dream, reverie, or vision" (1989, 80), Hayden talks of various ancient ritual activities including fasting, music, dancing and sleep deprivation that have been "widely used for inducing altered and ecstatic states of consciousness" (2003, 33). The latter also makes the interesting point that the "deepest religious experiences" have arisen from ritual stress conditions that hold much in common with times of climatic turbulence, including "starvation, thirst, physical exhaustion, sleep deprivation" (2003, 33). Of course, the big difference today concerns the justice issues of who is fuelling such numinous experiences of our climatically responsive world. If we are to find a sustainable and just way forward, I think it will have much to learn from the epistemological and ontological lessons that informed our ancestral ways of living and thinking. We are in need of an etiquette that deeply challenges the individual self-interest that has driven Western culture and the planet's economic globalization since the Enlightenment.

In considering the scale of climate and social uncertainty that is confronting our wobbling global consumer culture and the self-interested ethic that fuels it, I am drawn one last time to the paraductive thought of Thompson. He writes that "Great ages of transition, with their attendant changes of climate – as we ourselves seem about to discover – are difficult times of decreasing culture" (1981, 120). Though the preceding analysis suggests that even in times of deep uncertainty core parts of culture remain with us to be wisely recovered, I think what Thompson is referring to is a decreasing usefulness or energy behind prevalent cultural ideas and practices that have reified in their relation to ecology, climate and cosmology. That is, they have little capacity to sense the unknowable power their own life is based in, and thus the dominant cultural stories as currently configured become brittle and dry in their thought and application. Until the individually self-interested ethic that supports global consumerism, Canadian tar sands development, or the broader neo-liberal economy is sufficiently cracked open, it is highly unlikely we will wisely choose an ancestral etiquette that is needed to sustain human life in a climatically changing cosmos.

## References


Bateson, G. 1972. *Steps to an ecology of mind.* New York: Ballantine Books.

Bellah, R. N. 2011. *Religion in human evolution: From the Palaeolithic to the Axial Age.* Cambridge and London: Harvard University Press.

Burroughs, W. J. 2005. *Climate change in prehistory: The end of the reign of chaos.* Cambridge and New York: Cambridge University Press.

Calvin, W. H. 2002. *A brain for all seasons: Human evolution and abrupt climate change.* Chicago and London: The University of Chicago Press.

Carrothers, G., S. Kline, and J. Livingston. 1968/1988. *FESKIT: Essays into environmental studies: Being some interpretations and amplifications of the FES curriculum model.* Toronto: Faculty of Environmental Studies, York University.

Cheney, J. and A. Weston. 1999. Environmental ethics as environmental etiquette. *Environmental Ethics* 21: 115 – 134.

Deloria, V. 1994. *God is red: A native view of religion.* Colorado: North American Press.

Hayden, B. 2003. *Shamans, sorcerers and saints: A prehistory of religion.* Washington: Smithsonian Books.

Issar, A. S. and M. Zohar. 2004. *Climate change – Environment and civilization in the Middle East.* Berlin: Springer.

Leduc, T. B. 2010a. *Climate, culture, change: Inuit and Western dialogues with a warming North.* Ottawa: University of Ottawa Press.

Leduc, T. B. 2010b. The fallacy of environmental studies? An interdisciplinary foray thru Canada's academic programs. *Environments* 37 (2): 1 – 28.

Leopold, A. 1970[1949]. *A Sand County almanac: With essays on conservation from Round River.* New York: Ballantine Books.

Livingston, J. A. 2007[1981]. *The John A. Livingston reader: The fallacy of wildlife conservation and one cosmic instant.* Toronto: McClelland and Stewart.

Livingston, J. A. 1994. *Rogue primate: An exploration of human domestication.* Toronto: Key Porter Books.

Primavesi, A. 2009. *Gaia and climate change: A theology of gift events*, London and New York: Routledge.

Rappaport, R. A. 1999. *Ritual and religion in the making of humanity.* Cambridge and New York: Cambridge University Press.

Schönfeld, M. 2011. Introduction – Plan B: Global ethics on climate change. *Journal of Global Ethics* 7 (2): 129 – 136.

Shepard, P. 1996. *The only world we've got: A Paul Shepard reader.* San Francisco: Sierra Club Books.

Shepard, P. 1967. Ecology and man: A viewpoint. In *The subservice science: Essays toward an ecology of man*, edited by P. Shepard & D. McKinley. Boston: Houghton Mifflin.

Thompson, W. I. 2001. *Transforming history: A curriculum for cultural evolution.* Great Barrington, MA: Lindisfarne Books.

Thompson, W. I. 1989. *Imaginary landscape: Making worlds of myth and science.* New York: St. Martin's Press.

Thompson, W. I. 1981. *The time falling bodies take to light: Mythology, sexuality and the origins of culture.* New York: St. Martin's Press.

Wenzel, G. 1995. Ninqiqtuq: Inuit resource sharing and generalized reciprocity in Clyde River, Nunavut. *Arctic Anthropology* 32 (2): 43 – 60.

Wilson, E. O. 1984. *Biophilia.* Cambridge: Harvard University Press.

Wilson, I. 2001. *Before the flood: The biblical flood as a real event and how it changed the course of civilisation.* New York: St. Martin's Press.

Wood, P. 2010. Commentary: You say you want an environmental studies revolution. *Environments: A Journal of Interdisciplinary Studies* 37 (2): 43 – 46.

# Index

Page numbers followed by 'n' refer to notes